THE EDUMACATION BOOK

THE EDUMACATION BOOK

Amazing Cocktail-Party Science to Impress Your Friends

By Andy McElfresh

Foreword by Kevin Smith · Illustrations by Kelsey Dake

weldon**owen**

TABLE OF CONTENTS

FOREWORD

Andy McElfresh is the smartest man I know.

Granted, the bar is not high because I've spent most of my life standing next to Jason Mewes (the cinematic Jay to my Silent Bob). There are some great thinkers and mighty minds amongst the people I'm lucky enough to call my friends, but Andy isn't just smart—Andy is brainy. His knowledge is thick and veiny.

I met the smartest man I know in the least likely of places: *The Tonight Show with Jay Leno*. It was 2001 and my first time on the show, and I was promoting my flick *Jay and Silent Bob Strike Back*. After the segment, the producer of the show told me I was funny and had good chemistry with host Jay. She asked me if I would ever want to work for *The Tonight Show* as a correspondent in the field and I jumped at the offer. The lady who would change my life then said, "I know the perfect segment producer for you . . ."

The bits were called "Roadside Attractions." The pitch was simple: In the wake of 9/11, people were flying less and driving more. We were going to present a collection of side sites to see along the way. So when I told *The Tonight Show* I was going to be in upstate New York in a few weeks to do a college gig, they organized a small crew and sent them along with Andy to meet me for "Roadside" duty.

We barely knew each other but our chemistry was unmistakable. Andy's dry wit had me chuckling and laughing all day, as we visited the Herkimer Diamond Mines and the holiday lights of Onondaga Lake. I'd never had anyone write for me—the idea seemed weird to a guy who fancies himself a writer. But Andy was so funny, I actually looked forward to the blue cards he would hold up after he scribbled fresh jokes on them. While I was interviewing the proprietor of "The World's Smallest Church," I could glance over his shoulder and find Andy with a humorous talking point to keep the conversation going. Where I ended, Andy began.

But more than being fun during the shoot itself, Andy was excellent to be around between takes and off camera because he was FULL of interesting factoids and information. Most people are armed with information about their own lives and maybe some gossip. Andy was armed with more trivia than had ever been in any pursuit. "Cocktail-Party Science" he called it, but I described it more as Andy saying smart stuff in an engaging, easy-to-understand way while getting important ideas across to dumbbells like me.

Even more useful, I would smuggle Andy's erudition back into my own world and pass it off as my own, pretending I had an advanced education that I didn't. People would say, "You're brainier than you let on," not realizing that the extra smarts I was manifesting belonged to someone else. Andy's brain was making me look good.

Years after we'd met, I would become something of an avid podcaster. One day, *The Tonight Show with Jay Leno* went off the air, and *The Tonight Show* moved back to New York City. Andy was momentarily jobless for the first time in decades, so I suggested we record a podcast. Since he was the smartest person I knew, it'd be easy to build a show around him being smart and me being stupid while I strain to be smart. With pop science, could we, indeed, teach old dogs new tricks? I suggested we title the show *Edumacation*.

Recording *Edumacation* all these years has been a delight because not only am I laughing lots with (and sometimes at) my co-host, I get to actually learn along the way. I'm in my late 40s, where the only thing it seems you really learn anymore is that you're way closer to the grave than you were in your 20s and 30s. But when I record *Edumacation* with Andy, I come away wiser each time. If Andy had been a high school teacher he would've been everybody's fave, because if he knows how to impart information in a popular podcast with a sneaky sense of humor and plenty of painful puns (seriously—he's Attila the Pun), he could've easily been the Robin Williams we all would've stood on our desks for . . . until he eventually got fired.

I don't want to unnerve you, but now that Andy's written the book on *Edumacation*, it's like you're holding Andy's brain in your hands. And might I suggest that if you read something cool in this book and decide to pass it on to others, honor those borrowed smarts more than I ever did by crediting the Thinker who made it all possible: If someone asks "Where'd you hear that shit?" just confidently reply, "From a guy that used to work on *The Tonight Show* who Silent Bob thinks is really smart."

—*Kevin Smith*

INTRODUCTION

Scholastic, the company that has the audacity to advertise their books right in the classroom, is awesome. While some people say that everything you need to know in life you learn in kindergarten, I contend that everything you need to know *about a person* can be learned from what they ordered from the Scholastic Book Club in third grade.

A lot of it was guilt-free junk. By "guilt-free," I mean they were books, which every parent wanted their kid to be reading. And by "junk," I mean cheap pop-culture properties rushed to press, like the book version of *101 Dalmatians* or the Apollo 12 moon-mission picture book, complete with fold-out lunar map and candid shots of my heroes wearing polo shirts and Ray-Bans. There were plenty of science activity books that included "real" investigative equipment, like injection-molded plastic magnifying glasses that could not focus enough light to so much as warm up an ant, let alone burn it (much to the disappointment of sociopathic third graders everywhere).

But there were also magnifier boxes for catch-and-release bug examination, secret code windows (tinted cellophane that revealed messages in multicolored drawings), pocket-sized laminated knot-tying guides, and so much more. I wanted them all, under the pretense that they would open up a world of scientific inquiry, but really because I was a kid and I liked crap.

I was raised in a house where books were Books with a capital "B," and the only ones with pictures that were acceptable were nonfiction histories, or the *World Book Encyclopedia*. By the time I was in Mrs. Joray's third-grade class, I didn't need to be incentivized to pick up a book with value-added photos and crappy plastic gewgaws—I was halfway into the Hardy Boys collection. But oh, *I wanted that bug box*. Scanning the pages of that wonderful book-club catalog, I plugged numbers into my internal formula that weighed guilt, curiosity, and self-indulgence on one side of the equation, and tried to get it to add up to the likelihood that my mom would write a check for it on the other.

My best friend Matt went straight for *101 Dalmatians* and paid the $1.29 in cash. He was always throwing money around like that: the pencil holders in his flip-up desk had at least five erasers that he had bought in an *art store*, each one more exotic than the last. Tom McCleary got a book that featured eight awesome paper airplane designs, with pull-out pages that had dotted lines to show you how to fold them. (He would spend the next year carefully tracing those pages onto blank paper to keep the book intact.) But me? I needed the perfect book: fun to read, short entries, with no frivolous language on the cover like "Collect 'em, Trade 'em!"

And there it was, on page four of the Scholastic Book Club catalog, across from *The Elephant Joke Book*. Two lines of ad copy, plain text cover, and they didn't even bother to position the book cover at a jaunty angle: *Fun Facts*. Sixty knowledge-packed pages highlighting the bizarre and astonishing world around us. It forever changed my life.

Of course, it didn't get into too many specifics, opting more for punchy, Ripley's-style zingers. And whoever wrote that book sure knew how to send chills down my spine with italics. So while Margaret Mead was sweating it out making mats all day so she could learn what it meant to come of age in Samoa, I could skip right to the good parts like this: "In the South Seas, patients pay their doctor when they're well, *and stop paying them when they get sick!*" WHAAAAAAAAT?

Of course, being introduced to pithy factoids at such an early age can lead to personality flaws. One of the many reasons my camera operator on *The Tonight Show with Jay Leno* hated me was my constant nervous spewing of fun facts to fill the void of contempt between us. To some, this kind of behavior makes you some kind of cross between Cliff Clavin from *Cheers* and Mr. Know-It-All from *The Bullwinkle Show*. But where some people dismiss it as nervous chatter, there are people out there who *actually like it*. Not only do they get chills from my verbal italics, it also inspires their curiosity to go down whatever rabbit hole we're talking about at the time.

Enter Kevin Smith. And by that, I mean when Kevin came into my life, we discovered that easy flow of information made for a fun friendship. (I do not mean *actually* enter Kevin Smith—if you're going to give that a go, at least buy him dinner first.)

Kevin has always loved being amazed, a character trait that has only intensified since he became a regular marijuana user. And, like every great entertainer,

Kevin realized that we were having too much fun to keep it all to ourselves: We had to take our science-y banter to the masses. So, in 2014, we started doing the *Edumacation* podcast, with its four chambers of knowledge: The Sci (real science), The Fi (science in fiction and pop culture), The Why (where listeners ask Kevin questions about science that he is eminently unqualified to answer), and The Bye (our look at science in the news and current events).

After the first episode, we found it necessary to add a fifth chamber to this cow-heart of edutainment: EdumaCorrections. It turns out that—no matter the amount of research you do, no matter how steeped in the subject you might be—at some point you're going to make a mistake. Thanks to the Internet, listeners have many ways to tell me I'm an asshole for getting something wrong, but to my credit the podcast is usually 99 and 44/100th-percent good stuff, with the occasional (mostly pop-culture-related) mistakes. That's right, you can get the air pressure at sea level wrong, you can misidentify the protein in CRISPR gene editing, you can misquote Niels Bohr all you like, and nobody will care. But God forbid you sing the wrong lyrics to the *F Troop* theme or accidentally state the Cruciatus curse from *Harry Potter* is for killing and not torture, or else the Internet is up your ass faster than you can say "Hufflepuff."

At first, the shows were more formal. Like we were doing a show. I would actually write out my explanations instead of using bullet points. Kevin would actually listen. And we would need to edit the shows, since this was a rhythm we were not used to. But after a few episodes, *Edumacation* became more like organized versions of our regular conversations, with me trying to stay on target and Kevin pulling us down whatever rabbit hole presented itself.

Of course, I am not without blame in that regard. If I want to pass the ball to Kevin during the show, I need only to press play on some of the characters we have developed over the years, and Kevin will launch into some kind of odd discussion between that character and whoever pops into his head. Those include the Emo Goldfish (featured in the fishbowl on the cover), Whistly the toothless old Xenomorph, the Cavity Creeps, Kevin's parents, and my favorite, a generic asshole listener/Internet troll who hates Kevin for everything he's done, and for the fact that he didn't make *The Matrix*. There are many more odd characters who are lodged in Kevin's post-modern Thurber brain.

With more than 100 episodes behind us, it is clear that not all the facts that make it into Kevin's brain find purchase there. It could be how I explain things, but in all likelihood, it has more to do with the weed smoke, as evidenced by the frequent sparks of a lighter heard over his microphone (it does get intense in there). While I do not partake of the ganja, my exposure is excessive, so much so that during an early live show at The Comedy Store in West Hollywood, I took a pee test and failed.

So, let this book be Dumbledore's Pensieve for Kevin, a repository of fun facts, thises and thats, and other bits of knowledge I have thrown against his brain that "blew his hair back" and may or may not have stuck. And for you, dear reader, let this be a collection of those ideas, stories, and occasional facts I began accruing back in Mrs. Joray's class that you might find suitable for cocktail parties, car rides, and, at the very least, bathroom reading. I keep telling my editor that they should accompany the book with a cheap injection-molded plastic magnifying glass, but he just smiles and tells me to keep writing.

Enjoy!

—Andy McElfresh

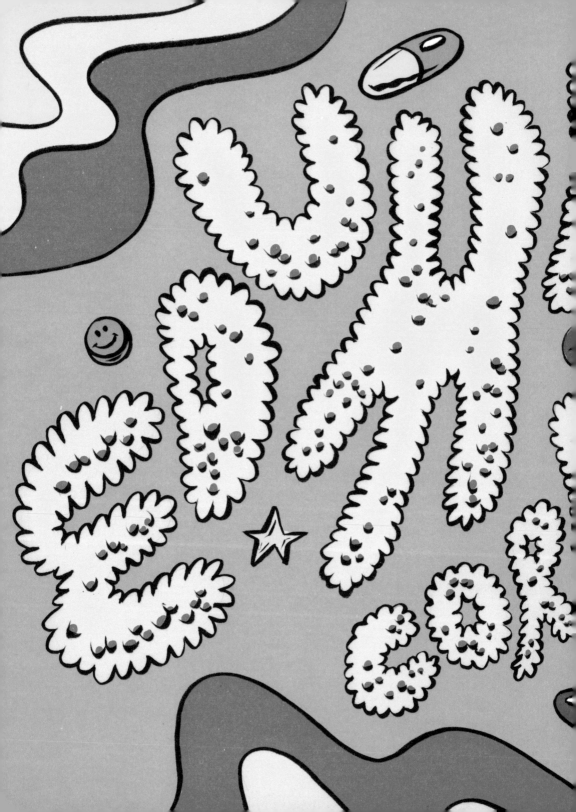

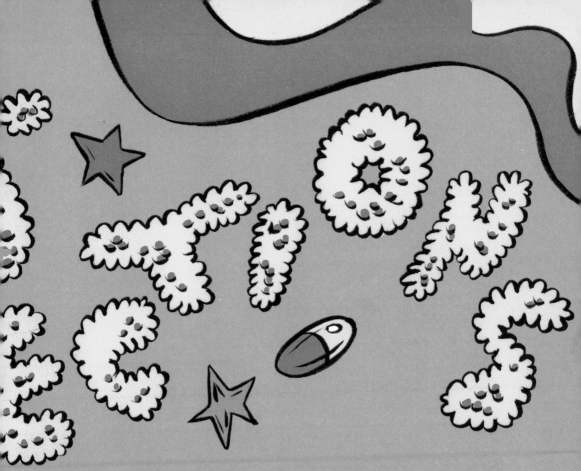

MANY YEARS AGO, I MADE DIRECTOR LEVEL AT MTV, AND VIACOM SENT ME TO MANAGEMENT TRAINING.

Everyone at work did whatever they could to avoid going, treating it like jury duty (not the Pauly Shore movie—nothing is worse than that). But it turned out to be two of the best days of education in my life.

Managing people is hard enough, but managing creative people is extremely challenging. You need them to put a piece of themselves into their work, give some of their heart and soul, and hand it over to you, and then you tell them what's wrong with it and hand it back. *And then they are expected to put even more heart and soul into it.*

It struck me then that, in a way, you would do well by thinking about the creatives the same way you would think about scientists. When scientists examine the body of evidence or make a discovery, they engage the creative brain to come up with a theory that fits the facts. Then they set about trying to prove their theory through more information gathering, fieldwork, or experimentation. But of course, scientists have it harder than the creatives do. Even though they engage those intuitive and creative juices, they must usually answer to the scientific community, as well as having to manage themselves.

The Kirk side of the scientist's brain wants to be proven right (just like in the writing field). It's much more interesting to work on your idea than someone else's (and it's more risky—a great way to get paid zero dollars per hour). But the scientist needs that Spock side to constantly reexamine what they are doing, regardless of whether doing that takes their beautiful sweetie of a theory and proves it wrong.

This creates a constantly evolving body of knowledge that can be everything from mundane to contentious, from trivial to profound. So, it makes perfect sense that the best that science can give us is going to change. And when a scientist makes a contribution, members of that community are encouraged to use it to advance their field of work. Were a creative person to incorporate someone else's work into his or her own, that would be plagiarism. But just as most patents begin with the words "Improvement on . . .," most scientific discovery has a built-in homage to the minds that came before. As Sir Isaac Newton said, "If I have seen further than others, it is by standing on the shoulders of giants." Which is crazy, because you would think that someone as smart as Newton would know there is no such thing as giants.

To the novice, it's easy to look back at how wrong we used to be, and how so-called corrections seem obvious to us now. But a more seasoned observer understands that we wouldn't have gotten where we are without the many missteps and enhancements to the body of scientific knowledge. And, to be honest, it's fun to point the finger at someone who was *so sure* they were right and call them an asshole from time to time.

So, get that pointing finger ready—and stay tuned for some mistakes that I will have to correct in the next book, because I'll be the asshole then!

Chapter 1

NOT SO FAST...

It's a given that scientists from the ancient world had some pretty terrible ideas about what was really going on in the natural world. But they also gave us a lot of the terms that we still use today.

Remember Hippocrates? Guy lived 500 years before Christ—I mean, before the common era—and was such a keen observer of the human body, he did a lot of naming. He identified and named several types of cancer, and because the tumors often grow in the shape of a crayfish (especially in breast tissue), which in Greek is called *karkinos*, that's where we get the terms *carcinoma*, *carcinogen*, and so on. When his works were translated into Latin by the Romans, they chose instead to use their word for "crab"–that's why cancer isn't just a medical condition, it's also an astrological sign. Even though cancer is very serious, I am glad we call it cancer and not crayfish (or, even worse, crawdad), so as to avoid exchanges like, "You serious?" "Serious as crawdad!"

Hippocrates is also famous for establishing the ethical standard for medical treatment. And I don't throw that word *famous* around lightly: Not only did he have a catchphrase, it's also named after him. "Do no harm" is at the core of the modern Hippocratic oath, which is not always easy to observe in modern medicine but is still part of the foundation of what we consider doctorin' in our society. (Not easy, because when you look back at some medical treatments—some of which are still in use today—you realize that doctors have a lot of flexibility in their interpretation of what harm is, and whether they are in fact doing it.)

An example of this is Lasik surgery. Yes, it is a body modification that improves vision by changing the position of the lens, but the way it's done is quite gross. First, the patient is made comfortable (they drop a Xanax on you, because you will stay awake while people are *cutting into an eye you can still see out of*). Then the patient's head is stabilized (read: strapped down to a table), the work area is exposed (your eye gets braced open like you're Alex in *A Clockwork Orange*), and the eye is numbed (with a variant of synthetic cocaine drops).

The team uses a microkeratome to make an incision across the cornea, which is peeled back, exposing the stroma (now they are basically scalping your eyeball), before a highly accurate laser removes corneal tissue, which changes the way light enters the eye, giving you better acuity. (They basically focus light and burn away some of your eyeball like a kid burning ants with a cheap, injection-molded plastic magnifying glass.)

I mention this because most of my eye doctor's business is in fixing botched Lasik surgery. Meanwhile, the guy who invented Lasik is in a five-year legal battle because his neighbors don't want him building a 50,000-square-foot wine cave directly under the Hollywood sign.

Back to the Hippocratic oath. The OGs (original Greeks) taking the oath were swearing to several ancient Greek healing gods, so it is obvious that things have changed—unless the current doctors are swearing to St. Raphael, patron saint of healing, or St. Michael, patron saint of the sick, or St. Jude, patron saint of lost causes, who isn't the most encouraging saint after whom to name a hospital: "Welcome to St. Jude's! We don't like your odds."

SPONTANEOUS GENERATION

Here is an idea that seems so absurd to us today, it's difficult to think that at one time, every educated person in the known world thought it was true. But they did, and if you were at a cocktail party and *didn't* think it was how the world worked, you were thought to be simply ignorant.

Spontaneous generation is the idea that life just kinda happens. Not in the Jeff Goldblum, *Jurassic Park* life-finds-a-way sense, or even in the "shit happens" kind of way—I mean, it literally *just happens*. No need to wonder where that wood louse, or a loggerhead turtle, or the howler monkey came from: All living things just sprang out of nothingness, ready to eat, digest, and make shit happen. And, to a large degree, we have maggots to thank for this.

Pliny the Elder (not to be confused with Pliny the Younger, or his imaginary grandson, Pliny the Tiny) was a naval commander, a writer, and a natural philosopher, and a personal friend of the emperor Vespasian. (His name has the same Latin root as the name of my beloved Vespa, which means "wasp.") Around AD 50, it seemed like anyone who took the time to make observations about the natural world and write them down was going to be taken seriously, but fame, power, and political influence played a role. Essentially, Pliny made shit up, and because he was who he was, everybody swallowed it.

His theory of spontaneous generation arose from the confusing, sudden, almost magical appearance of maggots, which are gross and highly noticeable. Why is it that you can take a piece of raw meat, put it in a closed container, and later find maggots feeding on it? Conclusion: Maggots just kind of appear out of nowhere, popping into their squirmy little existence without so much as a by-your-leave. Then they turn into flies, do a lot of walking on shit, and land on your food, then barf on it.

But Pliny took it further, applying the scant minutes of fly-maggot observation to the rest of the natural world. He wrote an account of a lamb that grew out of the ground. Forget the fact that people had been raising sheep for thousands of years and had seen them dropping out of lamb vaginas all over the place; when the emperor's buddy said lambs pop into existence out of a bush, everyone was like, "Holy shit, I didn't know that!" Such is the consequence of having a moron in power who doesn't know crap about science.

Finally, in 1668—which is an extremely long time later to buy into this kind of thing—Francesco Redi performed the meat-in-a-jar experiment with better controls. He did not allow flies to land on the meat before it was stored, and when no maggots grew, he proved that flies lay eggs. (I can't believe Francesco Redi is not a household name: Not only did he change the way the world thinks about maggots, he was born Redi.)

So, if someone tries to tell you animals spontaneously generate, they might be trying to lambush you. (And with that, you have learned that I will make any joke.)

PHRENOLOGY

In the early nineteenth century, a weird Viennese physician named Joseph Gall got an idea in his head about, well, heads. He had a theory that since the brain is the organ of the mind, it develops its different sections at varying rates in different people. Fair enough—they almost proved that with a 2009 study of *Tetris* in brain function. But then he started going off the rails, which is a nice way of saying that Gall fell into the trap of pushing a theory because it supported popular views. Like eugenics.

Gall postulated that you could judge not only the intelligence but also the moral character of people by examining their skulls. Aptitudes and tendencies could be accurately predicted just by examining the shape and topography of the patient's noggin. Again, just another scientist trying to do society a solid by fitting a theory to his observations. But various racial and ethnic groups share certain physical characteristics, so it was an easy jump to say that each of these groups had certain intelligence levels and moral values based on their lineage, *as dictated by the shape of their heads.*

Gall himself wasn't jumping to those conclusions, but others were. It wasn't until after Gall's death that this misguided theory really took off. It spread to America and France in the 1830s, and it seemed to come and go pretty fast. But then proponents of one movement from the United States took the show on the road to the United Kingdom. Known as the "phrenological Fowlers," named after fraternal phrenologists Orson Squire Fowler and Lorenzo Niles Fowler, they performed public skull readings and added their own layers of classification. They loved to rank different races and ethnic groups as "superior" and "inferior" and throw around terms like "master race."

While this is obviously bunk to nearly everybody now, phrenology was used to support other theories of eugenics (too awful to go into detail in a fun bathroom reader such as this), and while it gained and lost popularity and is now (supposedly) universally discredited, the British Phrenological Society was not disbanded until 1967. Luckily for everyone, it ended there, along with all ignorance and racial bias.

Oh, wait . . .

ALCHEMY

Remember when the first Harry Potter book came out and everyone was like, "Dude, you have to read this book—or, better yet, listen to it on tape." And you were like, "What's a sorcerer's stone?" And they were like, "Well, in the United Kingdom, where, until only recently, phrenology was an accepted theory, they call it *Harry Potter and the Philosopher's Stone*." And you were like, "What the fudge is a philosopher's stone?"—because you do not swear while having hypothetical conversations.

Well, it all has to do with alchemy. The philosopher's stone has been sought after for millennia; it is believed to be a substance that magically transmutes lead (and any other cheap metal, for that matter) into gold. It does other things, too—most notably, it produces the Elixir of Life, which, when consumed, makes the drinker immortal (I have a side theory that it tastes a lot like Diet Dr Pepper). With a philosopher's stone, you can become as rich as you want and live forever, so a lot of people have tried to find one over the years.

"Ha!" you say, "what kind of moron would buy into that kind of nonsense?" You start thinking of grinning village idiots and members of the Cheek-Biting Jackass squad. But would you be surprised to learn that one man, one of the most respected thinkers of all time, a scientist who invented the calculus and ruined the lives of high school juniors and seniors for all time, believed in alchemy and worked himself to death to discover the secret? (And, as an aside, came up with the whole idea of gravity?)

Yep, I'm talking about Sir Isaac Newton. When he wasn't standing on the shoulders of giants, he believed that there was a chemical process that would turn lead into gold. Which would make him rich. I mean, the guy understood gravity. (Not the Sandra Bullock movie—nobody did. Was George Clooney a space ghost?) This is crazy, on the order of a completely rational mathematician who spends all his money playing Powerball.

I guess it's about time I mentioned that Newton was a weirdo. He came up with the calculus but hid it in a trunk until someone else came up with it, then he pulled it out and said, "Fuck you! I came up with it first!" He also thought the Bible had a lot of secret messages in it that, when decoded, would tell him when the end of the world was coming.

It wasn't until Antoine Lavoisier came along in the late 1700s and put chemistry back on the rails that people started realizing that all this business about a philosopher's stone was just pissing up a rope.

STRESS CAUSES ULCERS

Sometimes bad science stays with us because someone's making money off of it.

Take ulcers. People with ulcers tend to have higher levels of stress than people without ulcers. So, naturally, stress causes ulcers, right?

Not so fast. In 1982, Drs. Barry Marshall and Robin Warren of Perth, Australia, went against the common thinking that no bacteria could live in the acid environment of the stomach. Marshall proved conclusively that many kinds of bacteria can indeed live in the stomach, and though they can survive the stomach's acids, the bacteria can be destroyed with conventional antibiotic treatments. Think of it this way: The traditional treatment for ulcers was the acidophilus bacteria in yogurt, and *acidophilus* means "acid-loving." C'mon, doctors!

Marshall presented his findings, and was promptly ignored.

One reason? Tagamet, a prescription stomach-acid reducer, was the most profitable drug of all time for what is now GlaxoSmithKline. It was one of the first drugs that revealed that it is much more profitable to treat an illness than cure it. Then Marshall came along and tried to screw with everything. (Please note that I'm not saying the big pharmaceutical companies are evil conglomerates bent on keeping us sick so they can make billions treating us, because they're big, evil conglomerates and I'm scared of them.)

It wasn't until Marshall voluntarily gave himself a peptic ulcer by swallowing a vial of the *H. pylori* bacterium—then curing it with a common antibiotic—that the scientific community came to realize that ulcer patients feel extra stressed because they have ulcers, not the other way around. And if you don't think scientists can be dedicated to helping mankind, just imagine what it must have been like proving your point by drinking a hot glass of salty ulcer juice.

WHAT DINOSAURS LOOKED LIKE

The producers of 2015's *Jurassic World* knew that dinosaurs had feathers, but they chose not to show that in the movie, because, as Steven Spielberg said, "People aren't scared of things with feathers." Really? Has he never tried to eat popcorn at the beach? Has he never seen *The Birds*? Wasn't Hitchcock his hero?

(To be fair, covering raptors in feathers might make them less Godzilla and more Liberace.)

Digging deeper, we find that scientists have known for a long time that dinosaurs had feathers. There was a movie in the early '90s that based a lot of dialogue on the idea that not only were birds descended from dinosaurs, they *are* dinosaurs. And what was that movie? The 1993 hit *Jurassic Park*.

Long-held perceptions are tough to change. I know that when I think dinosaur, I think bumpy lizard skin, not feathers. I also think brontosaurus—first discovered in 1879, which not only is what they called the long-necked protosaur in *The Flintstones*, it also means "thunder lizard," the most metal name of the Triassic era. This was an early dino-find, and they didn't get all the bones together in exactly the right way (as is sexually lampooned in the Cary Grant/Katherine Hepburn screwball comedy *Bringing Up Baby*), but it lasted as its own genus for more than twenty years, when it was reclassified as being in the same genus as the apatosaurus.

More fossil discoveries changed the classification even further, and in the 1980s it was decided that brontosaurus wasn't really even a separate species. Then a team of European paleontologists used statistical analysis and highly detailed scans of the differences in bone structure of the two beasts, and as of 2015, the brontosaurus is back, belonging to its own species *and* genus.

Which gives me hope for Pluto!

NEANDERTHALS WERE A DUMB, INCOMPATIBLE RACE

Let's get a few things straight about Neanderthals. When a fossilized skeleton was discovered in Germany in 1856, scientists named it after the nearby Neander Valley (*thal*, pronounced "tall," even though these dudes were short, means "valley"). They assumed, based on the small bodies and different head shapes, that Neanderthals were a different species. They are not. They are human, and in fact, even though the scientists didn't know it, these were the first early-human remains ever discovered.

Neanderthals are human because our ancestors had sex with them. Sure, that is not proof you are human, or else blow-up dolls would have driver's licenses. What I mean is, they successfully mated, and lots of us are running around with up to 4 percent Neanderthal DNA in our bodies. The facts that Neanderthals died out and that their fossilized remains were different enough from modern humans led scientists to erroneously conclude that Neanderthals were not human, when it is more likely the differences were more akin to racial classifications. More racism!

And they weren't dummies either. Neanderthals hunted, some of them were vegetarian—which may or may not speak to their intelligence—and they were the first humans to wear clothing. Based on that, I am assuming that Neanderthal women were the first to wait until their husbands were completely dressed and ready to go out before asking, "Are you going to wear *that?*" (and before sitting down to a dinner of woolly mammoth and asking, "Are you going to eat *that?*").

Based on the amount of junk in the human genome, acquired from viral infections, mutations that negated gene function, and environmental conditions that disable certain gene expression, that Neanderthal 4 percent could be responsible for a significant portion of what makes us—*Homo sapiens*—human. So don't go dissing Neanderthals. (Now, those Denisovans, that's another story.)

Chapter 2

URBAN LEGENDS

Urban legends come in many varieties. There are the conspiracy theories, like that the US government faked the Apollo moon landings—not just *Apollo 11*, but all six of the successful missions (and you get a bonus point if you can name an astronaut that landed on the moon who wasn't Neil Armstrong or Buzz Aldrin). There are the mystical-creature urban legends, like that of the Jersey Devil, who, by flying down chimneys and stealing babies, somehow managed to get a hockey team named after it. And there are the mysterious-death urban legends, like how Mikey, the kid from the Life cereal commercials, once ate Pop Rocks and drank Pepsi and was killed when his stomach ruptured.

But we did land on the moon. Jesse Ventura notwithstanding, all the so-called proof of a lunar lie is complete bunk. Look at it this way: For the hoax to work, *150,000 NASA employees would have to have been in on it.* And not blabbed. And every member of the White House staff would have known and kept that secret. Which is impossible—just look at the Trump administration.

And the Jersey Devil—the legendary creature for which the NHL team is named—was proven to be a hoax. Hoofprints attributed to it were later identified as having been made by a horse. Best of all, famed Philadelphia hoax artist Norman Jeffries admitted to staging a Jersey Devil sighting in the early 1900s by purchasing a kangaroo from a zoo and attaching claws and fake bat wings to it with glue. What a dick!

And we know that Mikey did not die from a fatal candy/soda combination. If fatal candy/soda combinations existed, Kevin Smith wouldn't have lasted long enough to make *Dogma*. Not only did I meet the real-life brothers who played the three kids in the Life commercial (we featured them in a segment on *Remote Control* at MTV), I even worked with a woman whose father was the food chemist who invented Pop Rocks. He was constantly asked why he would produce a candy so lethal, it would take a freckled child actor from us before his time. (But nobody ever asked him about Zotz or Bac-os, which he also invented.)

But there are some urban legends that are based on real life. Some are so weird and grisly, I'm having trouble coming up with fun, banter-y jokes about them. What follows are urban legends that have made the rounds and are based on true events.

THERE'S A DEAD BODY UNDER THE BED

You check into a motel just outside the city. Maybe you're trying to save a few bucks. Maybe you like the sound of families arguing, or couples having sex, or people having sex with prostitutes, or prostitutes arguing. In any event, you settle down on the bed to watch the free HBO (who doesn't have a friend's HBO Go log-in?) with a grape Fanta and an expired bag of Funyuns. After throwing back the hooker tarp—I mean the bedspread—you notice a foul odor. Like, a super-foul odor, something primal that hits you in the brain stem alongside the part permanently wired to make you run from alligators.

You frantically cast about in your head for explanations: Did I have Chipotle? No, this smells worse. Did I pick my nose and get Funyuns dust up there? No, this is still worse (but you know deep down that oppressive governments could use Funyuns dust for riot control). The initial shock begins to fade, and you realize the smell seems to be coming from *under the bed*. You don't want to get your bare feet dirty on the motel rug, so you peer over the side of the bed and . . . what a relief! It's one of those platform beds. Makes it hard to lose something under there. Back to HBO.

But the smell only gets worse. With a growing sense of dread and certainty, you slip into your shoes, get off the bed, and lift the mattress and box spring to find . . .

There's a dead body rotting under there! And it's covered with maggots that seem to have spontaneously generated!

Your frantic brain can't decide which is worse: Some poor schmo has lost his life, you're going to be up all night talking to the cops, or you just missed the last five minutes of *Game of Thrones*.

But don't be surprised, because it's happened before. *Lots of times.*

Urban legends often add details to make them seem real, like, "My math teacher's sister said . . .," but in this case, it not only happened, it also happened dozens of times, and continues to happen. In 1988 in New Jersey, and the following year in Virginia. In 1994 in Miami, in '96 in Pasadena, and so on. I'm surprised the motels don't have a written policy that charges you for dumping a body under the bed.

And the above scenario is nowhere near as gross as it actually gets: Most bodies are discovered several days into a customer's stay, meaning people are *sleeping just over a dead person's rotten remains!*

THAT'S NOT A DISPLAY—THAT'S A REAL DEAD BODY

I have always been scared of carnies. It's not irrational; I'll go as far as to say you should be, too. I once had a plan to take Kevin Smith to Winter Haven, Florida, to meet the cool folks at the Ringling Bros. and Barnum & Bailey Circus. We missed them due to scheduling, though, so we went to nearby Gibsonton, a community of former carnival workers. It's the only time I ever wanted to direct a segment from inside the van, because those folks are *scary.*

Maybe I have deep-seated fears. When I was six, my uncle Bill and his family took me, my brother and sister, and some cousins to the traveling carnival in Peaceful Valley, Pennsylvania. Uncle Bill was a state trooper, had done some awesome and dangerous things, and once let me fire his .38 snub-nosed Smith & Wesson. And he used to just blithely drop little comments that would make me lay awake at night.

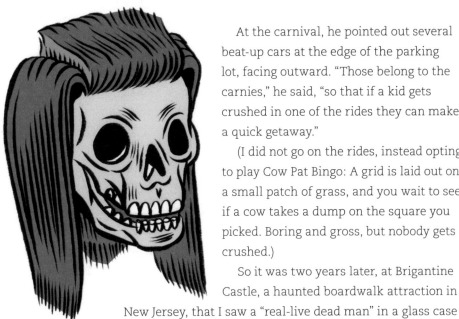

At the carnival, he pointed out several beat-up cars at the edge of the parking lot, facing outward. "Those belong to the carnies," he said, "so that if a kid gets crushed in one of the rides they can make a quick getaway."

(I did not go on the rides, instead opting to play Cow Pat Bingo: A grid is laid out on a small patch of grass, and you wait to see if a cow takes a dump on the square you picked. Boring and gross, but nobody gets crushed.)

So it was two years later, at Brigantine Castle, a haunted boardwalk attraction in New Jersey, that I saw a "real-live dead man" in a glass case on the second floor. And he turned out to really be dead. And I knew it. But I was eight years old by then and a man of the world.

Turns out this kind of thing was pretty common. Traveling carnivals and similar sideshows would routinely display dead bodies for money. Elmer McCurdy, the most amazingly appropriately named train robber the Old West had ever seen, was one such body that did a lot more traveling after he was dead than when he was alive.

McCurdy was fatally shot in 1911 in Oklahoma. After he was embalmed, the funeral director, angry that nobody was going to pay for it, decided to exhibit the remains as "The Embalmed Bandit." After he started to bring in some money, two men showed up and claimed to be McCurdy's long-lost brothers. They were actually James and Charles Patterson, *carnies*. They stole a dead body, as carnies do!

McCurdy changed hands a lot, played Jesse James in one show (with much more realism than Brad Pitt did in 2007), and was a prop in haunted houses and sideshows up until 1976, when his mummified remains were hung in a funhouse at an amusement park called The Pike in Long Beach, California.

When an episode of *The Six Million Dollar Man* was filmed there, a prop guy tried to move the body for a shot, accidentally ripped its arm off, and found mummified bones and other tissue inside. Doctors made the identification from a ticket stub found in McCurdy's mouth. He was buried on Boot Hill in Guthrie, Oklahoma. Due to his popularity, they covered him in two feet of concrete to keep the body from being stolen.

After all that time, he missed his chance to be on *The Six Million Dollar Man*. Bummer!

MAN DECAPITATED BY AN ELEVATOR

If you've been paying attention, you'll know by now that these are urban legends that are actually true, so I'm not going to phony up a scenario. It is worth noting that, just like Joey Tribbiani did on *Days of Our Lives* (which was totally fake, because *Days* is shot in Burbank, California, right next to where we filmed *The Tonight Show*, and Joey lives in New York), people fall down open elevator shafts all the time. But we're not talking about a simple fall here; we're talking clean decapitation.

Oddly enough, most of these elevator decapitations take place in hospitals. (And Joey played a doctor. Man, those *Friends* writers just weren't trying.) I researched why a hospital is so often the site of an elevator fatality, but my first instinct—that there is a button on those elevators that reads "Irony"—didn't pan out. Instead, it turns out hospital elevators are usually more robust, carry more weight, and work a little faster to accommodate emergencies.

In 2003, Dr. Hitoshi Nikaidoh got trapped in the closing elevator doors at Christus St. Joseph Hospital in Houston, Texas. The elevator hauled him upward, where his head was separated from his body, as a colleague stood by helplessly as the event seemed to unfold in slow motion. The head was left in the car, while the body was found at the bottom of the shaft. The ensuing investigation chalked it up to faulty wiring.

That same year, a patient in Louisiana was on a gurney, halfway out of the elevator, when it closed, shot upward, and crushed him against the ceiling, and there are several other cases of gurney fatalities due to malfunctioning elevators. The results are as though the elevator door frame becomes a giant, blunt guillotine blade.

In 1995, however, James Chenault distinguished himself by being not in a hospital, but rather in a Bronx, New York office building, when he unfortunately experienced a malfunctioning elevator. When it stopped between floors,

Chenault heroically kept the door open with his back as he helped passengers down to the floor. But the elevator suddenly lurched upward and decapitated him—and the remaining passengers rode up to the ninth floor with his head.

DEADLY WEDGIE

My mother was extremely talented in coming up with urban legends, even if they stopped with her kids because we knew they were utterly made up. She knew all the tricks, like starting the stories with variations of "When I was a kid" or tossing in details that were oddly similar to the behavior she was trying to discourage. Every story ended in death.

My favorite had to do with my obsession with remote-control model planes (dating from age five to . . . now). At the age of twelve, I considered myself ready for one. I had built hundreds of model planes, rockets, and boats, including string-guided planes that you flew by spinning in a circle until you threw up. I was a *Model Airplane News* subscriber. I belonged to the Radio Shack Battery of the Month Club (not directly applicable, but it did wonders for my confidence). Alas, it was not to be.

"When I was five, we went to the state fair," said Mom. "A man was demonstrating a remote-control plane when it turned on him. It flew straight at him, and no matter where he ran, it followed. *Then it chopped off his head.*" A quick search of remote-control-aircraft-related deaths in 1935 reveals that not only is this untrue, if it had happened, it would have killed its inventor and the sport would have died with him.

She also told my brother not to give me a wedgie, because it could kill me— this being the exception that proves the rule: She was actually right.

The victim of this brutal 2014 jimmy nut was a fifty-eight-year-old Oklahoma City man, well above the middle-school average age for this kind of thing. After a number of drinks, he got into an argument with his thirty-three-year-old stepson that turned into a multiroom fight. Then came the death blow: The younger man seized the older's BVDs by the label and pulled. Up. Hard.

This was no ordinary grundle flossing. The band was pulled over the head and around the neck in what is referred in the medical field as an *atomic wedgie*.

The man was strangled by his underpants. His stepson is doing thirty years in state prison, having to answer the age-old prison question "What are you in for?"

more than 300 times a day—and having to hear the ensuing punch line, "I hear it was crack," just as often.

THERE'S SOMEONE LIVING IN MY HOUSE

This horror is so common, Blumhouse Productions could make a feature about it for a few hundred grand: Weird things happen. Food disappears. Strange, creaky noises are heard at night. You feel like you're being watched—because you are! There's someone in the house!

Sure, it happened to David Letterman, but at least that woman had the courtesy to live there while he was gone and be under the delusion that they were married.

But in Denver in 1941, Theodore Edward Coneys, hard up and desperate, went to visit his old friend Philip Peters, who wasn't home. He let himself in anyway and found a tiny door to the attic. So, like people do, he squeezed in and lived there, stealing food and using the bathroom when Peters was out. One day, though, he was cooking in the kitchen when Peters surprised him, and Coneys killed him, slipped back into the attic, and hid from Peters's wife for months. Maids kept quitting, thinking the house was haunted, which it was—only by a live person. Police were baffled, so they staked the place out—and saw Coneys peek out a curtain. He died in the slammer in 1967.

We all have come to love Scranton, Pennsylvania, as the home to Dunder Mifflin Paper from *The Office*. But just down the Northeast Extension of the Pennsylvania Turnpike lurks the mysterious city of Wilkes-Barre. I would put in a pronunciation here but nobody knows how to pronounce it—the city is THAT mysterious. So it makes perfect sense that in 2008, while the Ferrance family was in the runup to another disappointing Christmas, they began to notice strange noises in the house. Christmas presents went missing. Police were baffled, which, if you google "Wilkes-Barre crimes," you will find is a pretty common occurrence. But when the family found footprints in a closet *that led nowhere*, they called the cops back in, and this time they brought dogs. They climbed into the attic to find 21-year-old Stanley Carter, who was wearing the mother's sweatshirt and the daughter's pants. The family did not ask for the return of the pants.

However, the Most Unobservant Homeowner of the Century award has to go to a Japanese man in the southern town of Kayusa in 2008. After food went missing from the fridge—*for a year*—the man leaped into action, installing a webcam that would alert him if there was movement. Instead of discovering, say, a trained otter that could open the fridge, take stuff out, close the fridge door again, eat his snack, and throw the garbage in the trash, he learned he had a fifty-eight-year-old homeless woman living on the top shelf of his closet. I mean, sure, she was small, but so are Japanese houses. The woman had somehow managed to get a teeny mattress up there and stay for a year without anyone seeing her.

To her credit, she used his shower regularly and was described as neat and clean. Which is evidently more than you can say for the homeowner, whose tiny Japanese house was cluttered enough that it was infested with another human. (Thank God he didn't call the exterminator. It's bad enough finding droppings in your cupboard. Can you imagine flicking on the flashlight and having a whole woman staring back at you?)

Chapter 3

FOOD - AND - DIETS

When my granddad went in for a checkup at the age of seventy-seven, his doctor told him to stop smoking cigars. Every cigar, he claimed, took a day off his life.

"Yes," replied my granddad, "but they take it from the end, when your life is shitty anyway."

I mention this because so much nutrition and diet news is aimed at longevity. Whenever someone becomes the oldest person in the world (an achievement, like becoming a king, that can be accomplished only through someone else dying), the first thing they ask is what the person eats: "What's your routine?" Recently, a 108-year-old woman in Florida claimed that beer, whiskey, and cigarettes were her secret. If that were the case, my granddad would have lived to be 5,000.

But this approach is behind much of the research into diet and nutrition, the only difference being that we take it seriously, and we flock in droves to whatever the latest trend is.

Case in point: Certain peoples in Peru have no heart disease or cancers. How do they do it? Could it be that, living in isolation high in the mountains and thereby having an isolated gene pool, the harsher environment weeded out people with those conditions earlier and made it impossible for them to pass their genes along to future generations?

Or could it be that they ate quinoa? Nutritionists, always on the lookout for "miracle grains," latched onto quinoa, and people are gobbling it down in massive quantities. Yet they're not losing weight. (Could it be that maybe the massive quantities have something to do with it?)

The worst part is, now that quinoa is so popular in the first world, the poor Peruvian farmers can't afford it anymore and have switched to cheap imported wheat and corn. Progress!

Different cultures have different approaches to nutrition. In the United States, we eat way too much of what we want, get a medical condition, treat it with drugs, and then have to limit our diet for the rest of our lives. In China, according to my acupuncturist, Dr. Jin (who once left a needle in my scalp, which I found out when I put my motorcycle helmet on and blood started dripping in my eyes), people are more sensitive to their bodies and eat well, and if they develop a condition, they try to treat it with food and herbs before resorting to medicine. And those people can make iPads, so there must be something to it.

What follows is a look at all of the contradictory claims about diet. Find the one that confuses you the least and stick with it!

VITAMINS

When my daughter, Daisy, was born by scheduled caesarean section, I took an audio tape recorder into the operating room to document it without having to see any gross stuff. Somehow, while the doctor was slicing, lifting, cutting, sewing, and weighing, the topic of conversation turned to vitamins. It turns out some doctors do *not* like vitamins.

And for good reason. In 2013, three studies were published that showed, from a very large sampling, that taking a daily vitamin has no effect on longevity or heart health, doesn't ward off memory loss and other aging-related issues, and has the same discernible effects as just taking a sugar pill.

Yet, it's a multibillion-dollar industry in the United States (so, if Big Vitamin comes after me, you'll know what happened). Sadly, the more health-conscious you are, the more likely you are to take a daily vitamin supplement. The "healthier" a store, the more vitamins. A walk through a Whole Foods market reveals many things: Grocery shopping is much more annoying when there's a deejay, people will pay four dollars for a pint of water with three spears of asparagus in it, and you have to look very closely at the meat labels to understand that's the per-pound price, not the total cost of the cut. It also shows nearly thirty yards of shelving filled with vitamins and nutritional supplements.

And there's another issue here. We will focus on gingko biloba, since it is touted as a wonder "drug" for memory and youthfulness. In 2015, the New York State attorney general's office investigated dozens of supplements sold in GNCs, Walmarts, grocery stores, and drugstores. They found that almost none of the brands that said they contained gingko biloba actually contained any gingko biloba. Some had ground-up flower petals, and to stimulate the making-you-feel-good effects of gingko biloba, some had caffeine in them. They extended the study and found this to also be true for echinacea, garlic pills, ginseng, saw palmetto, St. John's wort, and valerian root—most of the supplements lacked the very ingredient they claimed to provide.

A few even contained—but did not list as an ingredient—different kinds of beans to give them body, which can cause allergic reactions in some people. So, if you're taking a supplement that's been proven not to do anything for you, and it doesn't seem to be working, maybe you bought some magic beans.

Some supplements do work. A study of the *Gynostemma pentaphyllum* plant shows that, by steeping it as a tea and drinking it, one can lower blood lipids and cholesterol. When I was in my forties, I had a physical that revealed that not only had the drug companies lowered the guidelines for cholesterol, thus necessitating statin use, my own serum cholesterol had risen. My doctor told me nobody—*nobody*—ever lowered their cholesterol on their own, but since a friend of mine had turned to Eastern medicine for her daughter, I gave her herbalist a try. And it worked! I lowered my cholesterol in two months, passed the blood test, and forgot all about it, and now my cholesterol is probably roughly the consistency of a beef-tallow candle.

The Chinese tea use comes with other restrictions, like no caffeine and some other stimulants. And it works so well, Big Pharma is studying it to create a pill. But if making tea is too hard until that pill comes out, you can always buy the extracted supplements—if you like beans with your flower petals.

MEAT

Whenever a meat-centric diet comes along, people jump on it like Neanderthals on a woolly mammoth. I mean, I get that the members of the Lewis and Clark expedition nearly starved at one point, so they made up for it by eating 10 pounds of meat a day—each. They had a history, and had to deal.

But the plain fact is that Americans take in more than the 2,000 or so recommended daily calories, and we're fat because of it. A great nutritionist with no business sense would tell you that you need to limit your intake, limit red meat, add more vegetables, and get off the couch. But a great businessman with no nutritional sense would tell you to eat lots of the meat that you love, eat less of the other stuff, and buy his book. Like the guy who came up with the Paleo diet: Eat like a Neanderthal and you'll look and feel great, just like the Neanderthals did.

Well, by that logic, you would have to chase down, kill, clean, and cook everything you ate, which I'm imagining would take a lot more exercise than outlined in the many best-sellers. Given the choice between doing it the simple, easy way or the complicated, hard way, I think the Neanderthals would opt for binge-watching *Stranger Things* with a bag of Cheetos. Although that might be too scary—the image of Chester Cheeto on the package, that is, not the show.

The way the research works is tied to how insurance works: Scientists study risk groups—people with heart conditions, say—then dig into the subjects' habits and try to find where things went wrong. However, this does not account for the fact that people with heart conditions who eat a lot of meat may also belong to another common behavior group—they drink too much, maybe, or take in too many calories in general. Meat has been named as a culprit time after time, unless it hasn't.

Like in a recent Australian study, cited by Professor Clare Collins on the BBC Radio 5 *Live Science* podcast. It seemed to show that you can have 4 ounces of lean red meat a day, with a good mix of vegetables and fiber, and get more protein, zinc, potassium, and B vitamins than folks who don't. Just stay away from processed meats, like bacon, sausage, and the delicious lips and assholes of processed meat sticks.

Working on *The Awesome Show* for NBC was a great way to not only keep up with the latest in science, but to meet scientists and even sample some of their wares. That is why I can say with absolute certainty that lab-grown red meat is easily the grossest thing that I have ever put in my mouth—and I was a food writer so I have tasted some pretty gross shit. But now you can buy the Impossible Burger, which is a vegetarian patty that is infused with heme, the red stuff in your blood that tastes slightly metallic. They produce it with genetically modified yeast, then grind it up with some plants. It actually tastes better than the lab-grown "real" meat, but with a twenty-five-dollar price tag, it's not going to take the place of ground beef anytime soon. Because it tastes pretty good, but not twenty-five-dollars good.

You're going to have to rely on your common sense here, however. After all, you're not going to find that advice in any best-seller. Who's going to pay for that research? The non-meat, non-fructose corn syrup industry?

GMOS

GMOs, or genetically modified organisms, are not only a contentious subject, the term itself is also not easily defined.

Selective breeding is genetic modification. An ear of corn would look like the baby corn that Tom Hanks nibbles in *Big* were it not for thousands of years of selecting the biggest ears and breeding only their seeds. The same can be said for the British monarchy.

So, yeah, they're genetically modified, in that the Incans went ahead and did some unnatural selection. But the average grocery shopper, when scanning a label for "GMO" or "non-GMO," assumes that there was some lab that did some gene splicing, and the kale they're buying might have bits of frog in it. Just like the raptors in *Jurassic Park*. (I thought I'd make it a little farther before dropping another *Jurassic Park* reference, but there it is.)

However, the only reason we won't go all Malthusian and outpace our ability to grow food with our expanding population will be through genetic modification of the staple grains and vegetables. We need advances in food production to keep up with the medical advances that are keeping us going longer.

Complicating matters further, some GMOs (grains in particular) have been modified to be hardy under more difficult growing conditions. In 1985, PBS did a documentary called *Superseeds*, which detailed the efforts of Monsanto to produce grains that can withstand drought, cold, flooding, and so on. The grain they developed was tough to mill unless you roasted it in a hot-air popper. But, like kale, it was worth the extra effort, since it could alleviate famine conditions. (Wait—kale is never worth the extra effort. It should be burned.)

All good, right? Except the biggest issue with GMOs right now is going below the radar. Big grain companies are producing these superseeds, sure, but they are engineered so that the plants grown from them produce nongerminating seeds. That means farmers, who have set aside part of the harvest for next year's crops since time immemorial, now have to buy seeds every year. It also means that these companies have the patent to the genomes they have developed, and when plants grown from those seeds interbreed with plants on neighboring farms, *those* crops become subject to seizure.

And what about foods that are not genetically modified *per se*, but are the products of genetically modified organisms? Gene researcher (and all-around gene-ius) George Church, a professor at Harvard, is trying to end the long waiting lists for organ transplants by making pig organs compatible with humans. The first step was to cure a generation of pigs of PERV (porcine endogenous retrovirus), not only because it makes pigs less pervy, but because it's also the first step in making their organs transplantable. One side effect? The bacon, pork, and ham from these pigs *is healthier to eat*. I want some!

Look, according to WebMD, nine out of ten doctors think that GMOs won't do anybody any harm. A little bit of frog in your kale might make kale better. (No, wait—nothing will make kale better. It really is Satan's weed.) Besides, that one doctor doesn't recommend Trident. So, while the jury is still out, there are only two reasons to fear GMOs at this point: They can be used to control and limit seed supplies, and we just don't know enough about the long-term effects of eating manipulated foods.

SUGAR

If you're learning this here, that rock you're living under must have left a permanent mark: One of the purest things you can buy is sugar. Think gasoline is pure? Look at the octane rating: That's the percentage. And 91 is very high for gas.

Sugar, on the other hand, is as close to 100 percent pure as a product can be. Walter White couldn't make a purer product. I'm not talking about other sweeteners, like high-fructose corn syrup, just that sweet, sweet white sugar. Why does this matter? Because sugar has a lot of energy stored in it. In fact, it has *too much* energy. Call them kilojoules, call them calories—the lesson to learn from sugar is that not all calories are created equal. When you eat a piece of whole-grain bread, a certain percentage of the energy you get out of it is consumed by the energy required to break down the carbohydrates, proteins, fiber, and so on. The bulk of the grain and the dietary and soluble fibers help keep the system moving (I don't know why I'm using euphemisms at this point), vitamins and mineral are necessary for cellular activity and repair, and fats keep you making the hormones you need.

But with sugar, you just get the energy. Even if you are running a marathon, at the moment of eating pure sugar, your body will try to store that energy. It's just too much to burn off at once. And how does your body store energy? As blood lipids. Fats. Cholesterol.

I have a friend in Texas who spent most of his life on his family and career, and it took its toll on his health. So, after retiring early, he started to run every day, eating more veggies and really flipping a switch on taking care of himself. He is one of the rare folks who got down to his ideal weight and stayed there. And he followed his doctors' advice: Eat no oils and little meat, and when you run, chow down the carbs and eat energy bars so you don't run out of reserves.

He got diabetes.

His bloodwork was that of a man a hundred pounds heavier. And then he read a study about cutting out sugar in all its forms, and he got better.

High-fructose corn syrup is named as a villain because it's fake, it's cheap to make, and it's in a lot of products. They take the starch in corn and digest it for you with enzymes. And making glucose isn't enough, because it's not sweet enough: They process it further to produce fructose—found in fruit—which is much sweeter. But don't worry—Mexican Coke (with cane sugar) is just as bad for you, since, whether the food is artificial or natural, the human body is not designed to eat something with such concentrated energy.

Instead, give xylitol a try. Prunes are sweet because of it. And they make you go poo because you can't absorb xylitol. Dentists recommend xylitol gum because it reduces the amount of bacteria on your teeth. And it doesn't spike your blood sugar.

And if you read a study or article that claims sugar ain't so bad, check the source. It's almost always from an independent institute for the advancement of sugar.

Chapter 4

GOING MENTAL

Over the years, there have been a number of documentary series about UCLA's Brain Research Institute and the discoveries that have been made there. One approach that seems to offer many insights is when the team studies brains that have been injured or otherwise are malfunctioning. By studying the changes in cognition, mental and physical abilities, and personality, they can work backward to determine what the affected part of the brain normally does in healthy people. (Famed neurologist Oliver Sacks used this approach in his own explorations of brain function.)

There are so many wild examples that illustrate the complexity of the brain. One young man was injured in a car accident on a trip to Mexico, and while he had no other injuries, the bruise on his temple was enough for the doctors to keep him under observation. He called his parents and everything seemed to be fine—until they walked into his hospital room, whereupon he was overcome with distress. He was under the impression that these people were strangers, who only *looked* like his parents.

Doctors determined that there was damage to his temporal lobe, which affected his emotional response to sensory input. Hearing was unaffected, but everything he saw gave him no emotional response. When he didn't get a vibe from his folks, he figured they were phonies.

Something similar happened to my father when he was undergoing the advanced stages of Alzheimer's, combined with a form of dementia that caused parts of his brain to stop functioning and die off. He got pretty jumbled up— watching or reading the news was no problem, but the stories would become conflated with his own thoughts and memories, so, for example, he turned a story about illegal drug dumping in Europe to mean that he had been poisoned, thus his mental state. But as his temporal lobe became damaged, he was constantly seized by the feeling that everyone was an alien who had replaced the people he knew. Eventually, he thought that the nurses were swapping bodies every time they left his room.

Dad was a brilliant guy. And though he lost that gift, he never lost his curiosity, finding himself fascinated by what was happening to his mind and how it changed the world around him. By the end, it seemed like he was in a cartoon version of *Dark City*, with the walls and floors changing and moving around him, while people seemed to change and morph in interesting ways. The best way to entertain him? Tell him stories about himself, from the old days.

Like my father, I am fascinated with the strange ways the human mind can change reality. I am not unique for thinking in this way. From time immemorial, writers from Lewis Carroll to Philip K. Dick have drawn on these syndromes to tell stories. Here are a few examples that I particularly like.

ALICE IN WONDERLAND SYNDROME

No, it doesn't make you curiouser and curiouser, nor does it make you automatically follow orders when presented with a card that reads, "Eat me."

Micropsia, or Alice in Wonderland syndrome, is a neurological disease that affects the visual cortex and makes you see things much, much smaller than they really are. Your feet look like baby feet. Your legs look like dolls' legs. Your— well, I'll stop at the legs.

Some sufferers of micropsia describe it as like looking through the wrong end of a telescope. However, it seems to be more related to the viewer's perception of how far away objects are. So, while a building may be off in the distance, you may think it is close by, thus reducing its apparent size in your mind. And it's all in the mind: It has nothing to do with the ability of the eyes to see. In fact, better eyesight contributes to the eerie misconception of size, since your brain determines distance not only through binocular vision but also with other clues, like haze, blurriness, even the audio characteristics of the scene. One thing your eyes might do to contribute to the illusion is focus or aim at a point much closer than the object, while your brain rectifies the two images.

What causes it? I'll go from worst to best. Brain lesions, detached retina, macular degeneration, swelling of the eyes, Epstein-Barr virus (I've always wanted to open a bar called Epstein's, with the slowest service in the world), heavy use of hallucinatory drugs (like mescaline), epilepsy focused in the temporal lobe, or migraines.

Ahh, migraines. I can see how that would work, since I get them. They come in all varieties, but mine tend to be caused by histamine reactions, presenting in the top and back of the head and packing a mean wallop. Sometimes I'll get an ocular migraine, which doesn't hurt but messes with my vision. I was under deadline on *The Tonight Show* once, with four people in my office clamoring for a graphic I was kicking out to the control room with minutes to spare, and the whole time, the color of the room kept changing in my right eye from deep purple to a sort of bruised yellow. And I wasn't even on mescaline! (At least I don't think so, although I did order in from Baja Fresh.)

Psychological factors can play into this disorder as well. Patients with separation anxiety or feelings of inadequacy can, in a subconscious effort to overcome feelings of inadequacy, start to perceive other people as being tiny—a feeling that even sitting behind the desk in the Oval Office can't can't seem to fix.

So, if this happens to you, hold me closer, tiny dancer, and get a CAT scan!

ALIEN HAND SYNDROME

No, it's not a syndrome where a severed hand bursts out of your chest and skitters away into the bowels of your interstellar mining vessel. Alien hand syndrome is a condition that makes one of your arms appear to have a mind of its own. It will move involuntarily, grabbing hold of things or people against the will of the person to whom it is physically—though no longer mentally—attached.

This has to be a crazy kind of experience. As it was explained to Billy Bush, your arm will think it can grab a cat. Just because it's famous. It's a beautiful cat, and your arm just wants to grab it. (The crazy thing is that Billy Bush loses his job, and the arm gets elected president.)

While other parts of the brain can be involved, this problem seems to be rooted in the corpus callosum. That's the thin slice of fibrous nerve matter that connects the left and right cortical lobes of the brain. The most famous corpus callosum was that of Albert Einstein, remarkable in that it had a much higher amount of connectivity, which some neurologists think gave him more brain power and the urge to cry, "Vah-HOOO!"

It is thought that when this connection gets damaged, areas of the brain that have come to rely on communicating with the opposite side will send trained-in, complex nerve impulses to the affected hand and arm. It's like having Tourette's syndrome while trying to speak in sign language. While alien hand syndrome has popped up as a device in various stories, I refer you to Seth Green's early masterwork *Idle Hands* as an illustration. No, wait—*Evil Dead II*. Why do you think Ash had to chop off his hand? Oh, and it was one of the stranger things about the titular character in *Dr. Strangelove*.

So, what's it like to be in full control of most of your body while your arm is doing the blue act from Wayland and Madame's later work? In the *Journal of Neurology, Neurosurgery, and Psychiatry,* they describe the case of an eighty-one-year-old woman who had a minor stroke, after which her left hand kept punching her in the face and tried to strangle her.

Not all damage to the corpus callosum results in AHS. In the 1960s, a Philadelphia man was knifed through the center of the forehead, and afterward he was able to write and draw different things with his two hands at the same time. Based on the quality of the drawings, however, I submit that you don't need a brain injury to draw as crappily as that.

And the corpus callosum is occasionally cut intentionally in cases of extreme seizure disorders, though that practice is losing ground to medical and cannabidiol oil therapies.

GENITAL RETRACTION SYNDROME

This is a mental disorder that primarily occurs in men living in Asian countries, primarily India, China, and Japan. Known there as *koro*, it is the belief that your genitalia are retracting into your abdomen like you're teabagging liquid nitrogen. The common thinking is that it goes widely unreported, due to the feelings of shame associated with admitting that you think your junk is tunneling for your rib cage like it's Charles Bronson in *The Great Escape*. And remember, while the acute anxiety associated with this syndrome might make you more of a grower than a shower, it's all in the patients' heads. The ones on their necks, that is. They only *think* it's happening.

And it gets worse. Most sufferers become convinced that when their beans and franks go full turtle on them, not only will they be disfigured, they will also actually die. It's like the opposite of those plastic buttons that come in Butterball turkeys: When the button pops out, it's done. When your button pops in, you're done for.

This is not an isolated occurrence. Occasionally, there are outbreaks of "koro panics," in which members of a widespread group are overcome with the fear of losing their genitals. And while that doesn't seem rational, wait until you hear the reasons for what causes them. In China, men believe that female fox spirits—not to be confused with foxy female spirits—are the cause. In Singapore, Thailand, and other parts of Southeast Asia, it can be caused by unfounded fears of mass poisonings. And in Africa, sorcery is to blame.

Nowadays in Western culture, GRS is a pretty rare thing. But in the Middle Ages in Europe, it was widely held that a witch could cast a spell on you so that your junk could no longer be widely held. These accounts have led researchers to conclude that this is a universal condition. But before you start thinking that you're about to go full Jame Gumb from *Silence of the Lambs* (or Jay from *Clerks II*), know that treatments are available and there's nothing to be ashamed of.

In rare cases, some women can ideate that their breasts are shrinking into their chests, but the causes vary widely and are not related to the male fear.

FOREIGN ACCENT SYNDROME

First diagnosed in 1907, foreign accent syndrome causes patients to start speaking in a foreign accent. Sounds pretty straightforward, but having it is actually a little more nuanced.

While you might find yourself talkin' foreign, you can't actually speak a different language. Those who do either have parents who spoke it in the house growing up or they speak a pidgin version of that new language, based on limited exposure to it such as language classes or movies with subtitles. But that is *rare*.

What you actually get is an accent that isn't really an accent: Think the *Mission: Impossible* television series, or whatever the hell happened to Madonna after she moved to London.

It is usually caused by some kind of physical trauma, most commonly to the brain, though other physical or neurological events can cause it. For example, in 2016, a Texas woman woke up from jaw surgery to find that she had a generic British accent (which doesn't even exist, if you believe Professor Higgins in *My Fair Lady*).

Sometimes the accents are very good, like Gwyneth Paltrow's in *Emma,* and sometimes they're not so good, like Gwyneth Paltrow's in *Sliding Doors*. And sometimes they can be downright dangerous. In 1941 in Norway, a woman received a sharp blow to the head during an air raid and woke up with what sounded like a German accent. She was shunned by everyone (unlike Madonna, who got off a plane speaking a phony accent and should have been shunned by everyone).

Strangest yet, I can't find any cases where someone from another country got conked on the noggin and woke up with a neutral, Mid-Atlantic American accent. Or even a Canadian accent. There must be some part of the American brain that just wants to be from someplace else.

WALKING CORPSE SYNDROME

Cotard's delusion, or walking corpse syndrome, is the belief that you're still moving around and doing stuff, but you are actually dead. Think Larry King, but with—no, just think Larry King.

This is the very opposite of delusions of grandeur, and it's been diagnosed since 1788. Charles Bonnet, who was the first doctor to report on it, detailed the symptoms of an elderly woman who appears to have had a stroke. The paralysis was temporary, but when she was able to speak again, she insisted to her family that she was dead and that they need to shroud her up and stick her in a coffin. She complained the whole time. When she finally fell asleep, they dosed her with opium and put her to bed. This temporarily held off the delusions, but every few months or so, she would get back to griping about getting planted. Ironically, the only time they did bury her was when she was finally unable to ask for it.

More irony: Neurologist Jules Cotard got the syndrome named after him because of his articles about a French woman he treated in 1888. And while the syndrome describes people who are walking around thinking they are dead, this lady thought she was *immortal*. She also thought she had no insides: no intestines, heart, stomach, brain, or nerves. Her motto: No guts, no problem. Because she thought she was immortal and had no insides to feed anyway, she soon died of starvation, *on Cotard's watch!*

Interestingly, none of these people who believe they're the walking dead walk like they're in *The Walking Dead*. They just kind of act normally, save for the fact they feel they are empty vessels, automata who don't have the good sense to lay down and die.

And speaking of empty vessels that make people think their lives are going to be followed by an hour-long talk show hosted by Chris Hardwick: Again, it seems to come down to trauma, but it can be physical, mental, or emotional—and yet still arrive at these bizarre symptoms.

One more: In 2005, an Iranian man arrived at a hospital stating that he was not only dead, but also was now a dog, and that his wife was also a dead dog, and his kids were dead sheep. He had no family. He also claimed to be protected by God, which must've been a recent thing, given his whole imaginary family/death/being-turned-into-a-dog thing. This being Iran, they kindly administered electric shock treatment and sent him on his way.

Nothing like electroconvulsive shock treatment to really make you feel alive!

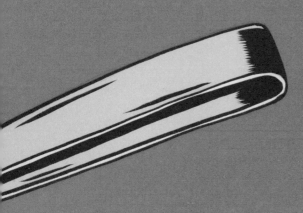

I HAVE BEEN INTERESTED IN SCIENCE, PARTICULARLY NATURAL HISTORY AND MEDICINE, SINCE I WAS VERY YOUNG.

My father was a medical doctor. He started his career as an anesthesiologist, but back in those days, the gas passer paid the highest insurance premiums in the OR, because if anything went wrong, they were always included in the lawsuit. So, when I was four years old, he went back to medical school and studied to become a psychiatrist.

My dad was a *very* capable man. And there is a Yankee "can-do" spirit that my dad—and his dad—instilled in me. Why hire someone to build you a garage when you've been noodling on it and have a clever way to do it yourself? *All by yourself.* That's how we roll in the McElfresh clan.

My grandfather was an illustrator who, suffering from partial alopecia that resulted in him losing all his hair at the age of sixteen, used his baldness to lie about his age and got a job as a fast-scribe artist at the *Philadelphia Bulletin.* They basically had funding and space for only one photograph on the front page, so the rest of the artwork consisted of illustrations scribed directly into the curved printing plate in the press room, minutes before press time. Backward.

By World War II, he was drawing all the war maps for the Associated Press. He would joke that the McElfresh name had appeared on more front pages of more newspapers than any other, and since he got a credit on those maps, it turned out to be true.

Granddad and Dad taught me that there was nothing you couldn't learn. Both men had an open dictionary on a stand next to his reading chair. Both men loved *National Geographic,* and would pore over the maps as they read the detailed travelogues. And both men were crazy about science, nature shows in particular, science shows in general, and, occasionally, Wonder Woman.

I think the pivotal conversation in my life was with my dad. I was a pre-med in college, and I had fulfilled all the requirements but one. And when I talked to Dad about following in his footsteps, he advised me to take a step back and really think about it. Going into medicine meant specializing—starting broad, then refining and closing off and becoming the best you could be within a field of study that would continue to narrow throughout your career.

He would have wanted nothing more than for me to become a doctor. But he also saw that there was too much to do, and too many things throughout the world for me to be interested in to systematically shut them out. So, he asked me what I would really want to do if that something was something other than being a doctor.

And I said, "Write."

So, now I get to read about anything that interests me, which is practically everything. And I get to talk about it with Kevin Smith, and got to make jokes about it for Jay Leno—and now, with this book, write about it for you.

Read the science section of your local newspaper. There's going to be something cool in there that impacts your life. But then go on to read it for forty years, and you find that a wonderful thing happens again and again. Everything *changes.* What we thought we knew about nutrition, for example, has flipped, flipped back, and flipped again. Science is all about formulating a theory, then pushing on it to see if it holds up—or breaks—then coming up with something else.

Even though I grew up during the Cold War, I have always believed that science makes our lives better, and can even make us better people. I say this here as a pre-apology for the poop-heavy facts that follow.

Chapter 5

AMAZING EVERYDAY OBJECTS

And now, a little more on my first job in New York, at the Zagat Survey (it's pronounced "zuh-GAT"). I should note here that I have to be careful what I say about the place because, before starting the company, Tim Zagat was the chief litigation counsel at Gulf + Western and Nina Zagat was a lawyer worth googling. The reason I bring Tim up here is that, when we got our first dot-matrix printer in 1986, he would stare at it whenever it was printing, mouth gone slack, and mutter, "That's goddamned amazing . . . goddamned amazing . . . ." The rest of us would take the opportunity to put a few more Renuzits in his office—for reasons I will not mention in print, since everyone in that office but me was a lawyer who had attended Harvard. (By the way, how do you know someone has attended Harvard? They tell you.)

This was not the only time Mr. Zagat exhibited this behavior. We got a laser printer, and he would stare at it. We got a Xerox machine, and he would stare at it. And when we moved to Columbus Circle, he got a pair of binoculars and would stare out the windows for hours, presumably at someone else's amazing printer.

Nowadays, we have so many more opportunities to marvel at everyday objects—so many that we rarely do. So, roll up those sleeves and get ready for some real cocktail party science right here. Let us begin.

DECK OF CARDS

There are 80,658,175,170,943,878,571,660,636,856,403,766,975,289,505,440,883, 277,824,000,000,000,000 possible ways to shuffle a standard deck of 52 cards. That's more than there are atoms on Earth. That means if you shuffle a deck of cards right now, odds are in your favor that you will produce a different sequence of cards than has ever existed in history—and *will* ever exist.. Some psychologists think our fascination with cards has to do with this randomness, and the rapidity with which the number of possible combinations decreases as each card is revealed.

While the idea of playing cards originated in China in the twelfth century, modern decks come from France, hence the weird royalty with knives in their heads and so forth. The kings represent real kings from history: spades,

King David; clubs, Alexander the Great; hearts, Charlemagne; and diamonds, Julius Caesar. People shuffle differently around the world: In Asia, they do the Hindu shuffle, which sounds like some kind of holy dance craze but is actually a method of pulling out cards from one hand into the other at random.

Many believe the deck is meant to reflect one year's time: fifty-two cards, one for every week. Four suits, one for each season. And if you count all the pips—the suit symbols on the cards—in a deck, you get 365, one for every day of the year. I guess the joker is for leap year.

I should note that when you play solitaire on your computer or smartphone and you pick Random Shuffle for the deal, you are not really getting randomness. The processor just can't deal with it. And the number of combinations you could possibly get are only in the hundreds of millions, not that crazy number above, which means that if you play Microsoft Solitaire on your phone for a few thousand years, you might end up playing the exact same game twice.

MICROWAVE OVEN

If you have a microwave (you do), then you have a marvel of technology. And if you have one with a dedicated popcorn button, congratulations: That could save you nearly 45 seconds over the course of your lifetime.

It's called a microwave because it uses very high-frequency (hence *micro*, meaning "super-small") radiation to vibrate the molecules in your food, which makes them rub together and heats them up. The heat comes from friction—not radiation—and despite what the conspiracy theorists say, it doesn't make your food radioactive (that's why microwaves are safe to use in speed-monitoring radar guns). Some molecules, like those in sugar, make better targets for the microwaves, so if you put an entirely frozen sundae in your Radarange, the sugar-rich fudge will get hot, while the ice cream will stay cold.

At the heart of your microwave is an inverter, which takes standard 120-volt AC wall current and turns it into very high voltage DC current, a property that means you can make your microwave an arc welder if you want. If you are a person like me, once you know this, pretty much all you want to do is turn your microwave into an arc welder, but your wife won't let you, because then you'd have to make your chili con queso on the stove.

You can also make an arc by putting metal in there, though some units, to keep ceramic plates from heating up, don't shoot microwaves at the bottom 1 inch above the oven floor.

Scalding from steam emitted by foods cooked in a microwave account for nearly a third of burns in household kitchens. And no one has ever accidentally cooked their cat in a microwave as a means of drying them off, but in 2014, a woman in the United Kingdom got fourteen months in jail for doing it on purpose.

PIANO

Maybe you don't have a piano in *your* home, but they're pretty damned near everyday items. The reason pianos are so heavy is that the part inside that makes the sound is a bunch of wires stretched across a frame that looks like a harp, and it has to be very strong. Each piano "string" has a tension of about 165 pounds, and multiplied by an average of 230 strings, that's 18 *tons* of tension. A concert grand has nearly 30 tons of tension, roughly the same amount you feel when meeting your partner's parents for the first time. I'm talking about a lot of tension here.

The word *piano* is just a nickname—it's short for *pianoforte*, which means "soft and loud" or "quiet and strong," which would make an excellent title for a Lifetime movie. When you press a piano key, a little felt-tipped hammer bangs on three or four strings, then backs off, allowing them to vibrate and make noise without damping. Hold the key down, and the damper stands off from the strings. Release it, and it goes back to home position, keeping those strings quiet. The sustain pedal lifts all the dampers from the strings, causing sympathetic harmonies on strings that weren't hit. And all of this first came together in about 1700.

Harpsichords have strings but use a quill to pluck them, which partly accounts for the sound difference. Harry S Truman was a concert-level pianist who once joked, "It was either politics or piano." Richard Nixon also played well, performing an original piece on *The Tonight Show* in 1963 after joking that "Republicans don't want another piano player in the White House."

On a personal note, Mrs. Talone, my piano teacher when I was six, was super mean and would whack me on the knuckles with a ruler. If I didn't practice, there was usually about 18 tons of tension leading up to my lesson. Sitting at a piano still gives me stress.

TELEVISION

What is the one thing you see every day and completely ignore? No, I'm not talking about the Conan O'Brien show; I'm talking about your television set. In many homes, a new, 52-inch LCD flat screen is the biggest consumer of energy, running at 400 watts on average. If you want to compare that with another electric thing, turn on a 100-watt bulb, and realize this thing uses four times as much energy. Then touch the bulb, and you'll realize why the Easy-Bake

Oven used one instead of heating elements. (I have to comment on the Easy-Bake Oven for a second: Why would you buy your—per the ads—daughter what was essentially a box of sheet metal that baked extremely tiny cakes with a lightbulb? I have no answer for this—it is rhetorical.)

Unlike a plasma TV, which has little fluorescent pixels that each emit light (making bad pixels more common), your LCD screen has a big, fluorescent light-emitting screen consisting of gas in strips (excited by electrons in the same way fluorescent bulbs in your garage are) behind a wall of diodes that polarize that light, turning it into different colors. So, they're not emitting the light, they are taking super-bright white light and blocking out everything but the color that the pixel is meant to be. Seem wasteful? It sure is.

But it's the same technology as your laptop screen, which can be more useful. I had an old Dell laptop that went on the fritz (motherboard, not worth fixing), and I removed the fluorescent panel and turned it into a light table for my daughter (then I went Mac and never went back). And the screen was still useable, and, being a Dell, the ribbon went right into an adapter I had and you could see right through it. If you want a taste of that sweet, sweet nerd-mod action, go ahead and google "see-through laptop screens," and you'll find folks who have modded their laptops so they're like looking through windows (even if they're running OS X).

YOUR SMARTPHONE

Unless you're Guy LaPointe in *Yoga Hosers*, chances are you have a smartphone in your purse or pocket or strapped to your arm as you jog down the sidewalk. To paraphrase Louis C. K., that thing is a fucking miracle—too miraculous to cover everything here.

The most oft-cited fact about iPhones is that they are so much better and faster than and have more memory and processing speed than the NASA computers used for the moon missions, or even the space shuttles. But, to be fair, you can say the same about a USB drive, so get over it. Smartphones contain more than a third of the elements on the periodic table. The processor alone has silicon (natch), phosphorus (discovered in pee), antimony (ancient Roman eyeshadow), arsenic (found in abundance in Cabot Cove), boron (rocket fuel igniter), indium (component of durable mirrors), and gallium (used in laser pointers).

The iPhone is the most profitable single product of all time, with more than 700 million sold. Of course, I don't have the patent-infringement data to back that up, but I'm going with it. And that doesn't include the (possibly 200 million) gray market iPhones available in Asia ("gray market" refers to products that the factory produces in surplus without telling the company that owns the rights to the product and sells on the side as the real deal).

I mentioned patent infringement, and you have to realize how hard it is to enforce that kind of thing in an everyday object that sophisticated—especially since it is cobbled together from many other successful engineered elements. Samsung, Apple's bitter, bitter rival, actually makes the Ax processors that power the iPhone, so every bit of code associated with running that screen gets reviewed by lawyers every time there is a software update.

The most expensive part of the iPhone is the retina display, which is truly incredible for both its resolution and the fact that it uses Gorilla Glass, which has been around for decades without much application and now has become the part of the iPhone that is in the greatest demand and often causes the most delays to production.

Finally, in true Steve Jobs fashion, the name of the freakin' device was not available, since someone had already thought of it. When Apple debuted it, "iPhone" was already taken by Cisco Systems. But Apple went ahead and used it anyway, preferring to not ask permission now and then apologize (in the form of answering a lawsuit) later.

Chapter 6

ALTERNATIVE MEDICINE

In Western medicine, people have always had a bond of trust in doctors. All of that schooling, the lack of personality, the authoritative air of someone in a lab coat who really knows how to wash his hands—it all combined to put the patient in a baffled-yet-reassured position as the doctor killed a goat and examined its liver, applied leeches to your forehead, or explained that the CAT scan wasn't covered by your insurance. I'm mixing up eras here, but to make a point: Up until recently, a doctor was a doctor, no matter the tech.

That was then. Nowadays, we have the Internet to give us a little bit of knowledge, terrifying stories of malpractice, and wrongheaded (or just wrong) fake news about the best way to stay healthy. We don't trust our doctors as much as we did, but it took a lot of trust to put yourself in the hands of someone who had the power of life and death over you. It's almost ironic that the better care gets, the more options we want and the more suspicious of doctors we are.

Imagine the courage it took back in the day, before sterilization and anesthetics and antibiotics, to walk into a doctor's office with an injury that might result in him removing your foot, wrapping your calf muscle around the remaining bone, and sealing it all up with a hot iron. Or consider the prospect of lying back to give birth and having the doctor provide a wooden spoon for you to bite down on. That was the extent of medical "technology" at one time, and it was the best that doctors could offer.

It's not like those days are entirely over: While doctors used to make with some pretty insane treatments, some of them are still in use today. God forbid you're in a terrible car accident and you get rushed into surgery. The doctors and staff in the ER surgery are going to treat you almost exactly the same way as if it were Civil War times—and I'm not talking Marvel's *Civil War*. The gear might be cleaner, the diagnostic equipment more informative, the knives sharper, and the painkillers more effective, but essentially, they're going to get in there and start hacking away the damaged parts and sew you up with needle and thread.

On the other hand, there are simply some questionable things happening on the fringe of mainstream medical treatments. At first, when I began writing this section, I was concerned that by including quack practices, medieval methods, and low-tech treatments, I was going to encourage Gwyneth Paltrow to swoop in and include them all on her Goop website. Turns out some of them are already there, which is both a horrifying admission that I went on that site and an implication that these outdated medical treatments belong in the world of vagina rocks and foreskin face creams. Enjoy!

CLEAN YOUR WOUNDS—THE MAGGOT WAY

Maggots (the larval stage of flies) usually appear on rotten organic matter and are particularly fond of blood and meat. So, rather than wait for a fly to deposit its eggs on a festering wound, doctors would apply clean maggots to the area and let them do what they do best: Eat the gross stuff. And surprisingly (at least to me), the maggots leave the healthy tissue alone.

When they're done, the maggots are scooped away and killed, which I think is a bit harsh. After all, they did the work—now they have to die? Maybe the thought is that once maggots have had a taste of human flesh, they can never go back. More likely, the doctors using maggots can't believe what they've just done and want to erase the whole thing from their memory forever (or at least until the next dirty wound comes along).

My friend David, who interned in the emergency room at Manhattan's Bellevue Hospital, has seen many strange and horrible things. And the hospital has responses for each and every one of them—like, if a man arrives with a 3-pound eggplant in his rectum, the correct procedure is to remove it, then tell him to chew his food more carefully. And it is not uncommon for the homeless to arrive with maggots nibbling at their feet, particularly when the weather is wet and the dead skin in their boots and shoes gives off the stench of rot. The protocol is to first reassure the patient that, yes, they are not hallucinating, and they actually do have maggots on their feet (although the radioactive genius spiders aren't real). Then, the young doctors kill the maggots by wrapping the foot in a plastic bag and tossing in a few cotton balls soaked in ether.

My friend said that it was not uncommon to perform this procedure a few times a week at certain times of the year, and in addition to having grateful patients, you've never seen cleaner feet.

URINE THERAPY

While "urine therapy" shares four consecutive syllables with HBO's *In Therapy*, it is much more shocking and dramatic.

If you're at the beach and get a jellyfish sting, some jackass is going to insist that the best thing for the wound is to pee on it. There's a scientific explanation behind this: That guy wants to pee on you. As for helping the sting, it simply doesn't work. However, it's not the only time that some medical whiz decided to use medicinal whiz to cure what ails you. At different times, urine has been used as a cure-all for your skin and hair. You would drink urine to cure Victorian maladies like the vapors and the spinster's complaint.

You heard me right: Drink your own urine. There's a long history of it, which makes sense for cavemen: You know, what else is there to do until the wheel comes along? In ancient Rome, urine was used as a mouthwash for whitening the teeth (which might make sense, as pee-pee has trace amounts of phosphorus in it). If you were rich, you could buy the urine of a badass like a gladiator, even though there was no evidence that gladiator piss made your teeth whiter than regular old piss from a slave. (If I was an ancient Roman marketing guy, I would have called gladiator piss "G-whiz.")

There is a diagnosed mental condition where people become obsessed with drinking urine, even believing that drinking the pee-pee of the young will keep you from aging. In fact, urine is sterile and is usually safe to drink—and by "usually," I mean it doesn't have a lot of bacteria in it, though the body will often secrete heavy metals and drugs in the urine without breaking them down.

A great example of this explains one of our culture's most popular myths: Laplanders drink reindeer pee to have hallucinations (which is where we get our flying-reindeer lore). Once you've had it, you can drink your own pee and get high again, or pass it (pun intended) along to your friends.

If you're dying of thirst, you can drink your pee for hydration. But you have to drink it: The only time peeing on your skin for any kind of remedy or hydration comes in the form of treating athlete's foot, but that has never been proven and just seems like an excuse to pee in the shower.

Now, if you want to turn the tables and get your doctor to down some piss, you don't have to wait for him to be in a Russian video with a president. Just tell him to do it old-school: Until the twentieth century, diabetes was often diagnosed by tasting the patient's urine to see if it was sweet. Which has got to taste better than gladiator pee.

BLOODLETTING

Until the nineteenth century, the old medical belief in the four humours prevailed. That is, your body is filled with four humours, or fluids: blood, phlegm, yellow bile, and black bile. In spite of the fact that humans have been either delicately dissected or violently chopped open for millennia, the medical establishment thought it was true, so, for all practical purposes, it was.

Among this selection of grossness, blood was believed to be overproduced, throwing the body out of balance. Symptoms included impertinence, hysteria, liverishness, malaise, ennui—pretty much all the symptoms that could be taken care of with some Girl Scout Samoas and a 5-hour Energy drink.

To remedy this, physicians would open a vein and take out some blood using a lancet (a word that is still the name of a modern-day British medical journal). Even more serious conditions, from irritability to nervousness to insomnia, showed marked signs of improvement after a supervised bloodletting. And why not? Who has the strength to stay conscious when they're a few quarts low? And you thought it was inconvenient taking a daily aspirin or Lipitor. Bloodletting was the anti-Ambien (see Chapter 11): Instead of unconsciously getting up in the middle of the night and eating an entire cheesecake, you might not get up at all.

Once medical science figured out that taking blood out of someone helps only if you actually stick it back into someone else who needs it, though, bloodletting died out as a treatment.

There are exceptions to this: Polycythemia (having an abnormally high red blood cell count) and hemochromatosis (having too much iron in the blood) are ameliorated by the letting of blood. Or you can just drink tea that's been steeped a long time: Tannins in the tea will drop your iron levels in no time.

Incidentally, the bloodletting process was pretty straightforward, with the doctor opening a vein in your arm and letting it drip into a collection jar. But there are folks who faint at the sight of blood, so they require a screen, or for the lancet to be applied where they can't easily see it. Researchers believe this fainting mechanism is actually part of the genetic plan: When you see your own blood, the sudden automatic drop in blood pressure that induces fainting will make you bleed less quickly, thus minimizing blood loss before it has the chance to clot.

TREPANATION

The expression "I need this like I need a hole in the head" probably refers to the practice of trepanation, or drilling or cutting a hole in your skull. I refer, dear reader, to the film *Master and Commander,* wherein Dr. Maturin saves the life of an old sailor with "Hold fast" tattooed on his knuckles who is dying from a depression fracture to the skull. He opens a flap in the skin in such a way as to allow the blood to continue to flow, then removes the offending patch of skull and has the ship's armorer fashion a silver coin into a patch that can be nailed to the bone with a silver nail. While the retelling of the event by the sailors later in the film is pure exaggeration, the depiction of the trepanation is extremely accurate to the time.

Which is to say it is accurate to *our* time, save the sanitary conditions and anesthetic. I have an uncle with a plate in his head from Korea. (The Korean War, that is; he didn't order a Korean plate and have it installed years later.)

Though there are some practical reasons for doing this, like removing crushed bone that presents a danger to the brain, cultures like the ancient Mayans liked to treat mental illness by drilling a hole in your head to let the evil spirits out. While this might seem like an extreme hangover cure, there is some merit to the practice, and the Mayans were no strangers to madness caused by pressure in the skull: The members of the infamous Conehead cult were a roving band of crazy priests who would kidnap young boys and bind their heads to reshape them into cones, making them crazy, and so the cycle continued.

Certain skull injuries are treated by removing the depressed fracture with a trepanning saw, a nasty device straight out of *Dead Ringers*. In the case of some head injuries in which the patient is unresponsive, emergency doctors may decide to drill a hole in the skull to relieve intracranial pressure, and even though these are only in extreme cases, some New York hospitals report treating injured homeless men with no medical histories and finding two or even three holes drilled behind their ears.

Surprisingly, we may all have holes in our skulls at some point in our lives: Brain interfaces are being developed to aid in memory and brain function. And optogenetics—the controlling of brain function and gene expression with tiny lights—would require that you get that extra head hole, for the wires. And based on the miasma of wires behind my desk, I'm thinking there's a fortune to be made in medical detanglers.

VAGINAL MASSAGE

I get it—I said "vaginal." But this was a real treatment, offered first in that most dubious of medical periods, Victorian-era England.

The theory is simple: Were a woman to become "hysterical" in the personal sense of feeling excitable, not the public sense of having a breakout Comedy Central show, her doctor might prescribe a course of getting a rubdown in her na-na. Any improvement in the well-being of the patient would be ascribed to this, so the doctor would continue the practice, as often as daily, for as long as he saw fit, or at least until she caught him grinning fiendishly under his handlebar moustache.

Digital manipulation of a women's nethers was not just limited to cases of hysteria. In the late 1800s, a one Dr. Swift advertised in woman's magazines for his gifts at treating "all disease of the mid-quarters from neck to knee," through the art of "fine gentle massage." The picture in the ads depicts a dude in a morning coat, down on one knee, making a grab for a lady's asshole neighbor right under her petticoats. She demurely looks away, but the look in old Dr. Swift's eyes is something akin to Harpo Marx when he's about to start chasing a woman around the lobby of a hotel.

For a pop-culture reference, see *The Road to Wellville*, the film adaptation of T. C. Boyle's book that detailed many of the quack practices of the Kellogg brothers in their sanitarium in Battle Creek, Michigan. Among their other treatments: wrapping patients in freezing towels and leaving them outside to lower their body temperature to near-hypothermic levels, bathing guests in the penetrating heat of the newly created electric light, and restricting their subjects to a yogurt-only diet—and if that wasn't bad enough, the regimen included frequent yogurt enemas. Jamie Lee Curtis would be proud.

Book, movie, and story end when a man, already fed up with his wife's devotion to this quackery, discovers her outside by a creek with a doctor who is letting his fingers do the walking in her flossy dune of femininity. After a *very* harsh "I bid you good day, sir!" they return to their normal lives.

But the story isn't over. Nowadays, getting a prosciutto shiatsu is getting more and more common. On the bad side, some college athletic trainers have been accused of taking advantage of young women by treating back pain by putting a coat of polish on the douche canoe. But of late, many reputable magazines have reported that women are having more satisfying sex by letting an expert knead their ladyparts.

WARNING: There are many words in this paragraph that will make it seem like I made this stuff up, but, at least according to a 2016 article in *Women's Health* magazine, it's all true. Isis Phoenix (!) is a somatic sexologist (!!) who ascribes to the tantric tradition (!!!) of a yoni massage (!!!!). *Yoni*, Sanskrit for "vulva," means "sacred portal." A yoni massage is a ceremony where a woman is invited to "seize touch on her vulva . . . to cleanse a sense of energy."

Whew! I'm glad vagina massage has made it out of the dark Victorian age and into the light of Science!

Chapter 7

REAL SUPERPOWERS

Comic books have long depicted the fantasy of possessing speed, strength, and other powers that would set you apart from your fellow humans. (They also envisioned a world where spandex existed long before DuPont rolled it out.) But super-abilities are just as real to some people out there: Through innate ability, chemical enhancement, or incredible discipline, they prove that superpowers *do* exist.

I want to be clear here: There are people out there with powers that approach the super. Maybe not *X-Men* super, but certainly *Alpha Flight* super, or at least *Kick-Ass* super. And what is truly amazing about them is that unlike all the hundreds of secret identities created over the years, they are famous, or at least Internet famous, or in some cases local-news famous. Let us celebrate them for being the super-people they are.

That being said, if archvillains also exist, I hope that we would try a diplomatic solution before we sic these badasses on them. Except maybe the eyeball guy (see below).

THE CALCULATOR

One of my favorite villains on TV's *Arrow* is the Calculator. Somehow, he can use his brain like it's IBM's Watson, and yet he comes up with a lame villain name like the Calculator. What, "the Slide Rule" was taken? I guess if your power is to be able to calculate the tip on a six-top check, you can get away with it.

The real-life women in *Hidden Figures* worked as human computers in the Mercury and Apollo space programs and were instrumental in helping the United States win the space race. But they couldn't hold a roman candle to the queen of cypherin': Shakuntala Devi, from Bangalore, India. She could multiply two thirteen-digit numbers, in her head, in under twenty-six seconds—which includes her reciting the twenty-six-digit answer. When she was a teenager, her father had her performing in the circus, then later toured her on his own, unaware that she could calculate down to a hundred decimal points how much her dad was exploiting her.

Another skill? Give her a date from history and she can tell you the day of the week, although that's not unique—my nephew Timmy can actually do that, too. (It's not the most amazing thing he can do, but it's crazy fun.)

DARKMAN

In Sam Raimi's *Darkman*, Liam Neeson's character drew strength from his inability to feel pain—and his ability to art-direct the set when he got mad. The condition is called analgesia, like what that creepy guy in *The Girl Who Played with Fire* has. That film showed some of the drawbacks: Lisbeth Salander's older brother, that Rutger Hauer–looking dude, cannot feel pain, and the movie illustrates how if you had analgesia, you'd have to check your arms and legs for cuts so you didn't bleed all over the Ikea flokati when you got home from being a thug.

A lot of people are afraid of pain, but this idea frightens me even more: Imagine that you're swimming at the beach and cut your foot on a razor-sharp shell. Most of us would howl in pain and do what we could to stanch the blood. But someone with analgesia could go for an hour before realizing they just gave a creepy lunch of a couple of pints of blood to a hundred sand crabs and will need (another) tetanus shot.

Analgesia is a real thing, and Kevin and I have met someone with it. Tim Kridland plies his trade as the Human Pincushion in various sideshows and TV appearances. When we met him at the Seaside Heights Boardwalk sideshow, he explained that he could literally take out his own appendix without drugs if he had the medical know-how. Unfortunately, his skills were limited to piercing his skin with huge needles. When he jammed a knitting needle all the way through his upper arm, Kevin and I had the proper reaction, which was to go outside and vomit over the rail.

DOCTOR MID-NITE

DC Comics had a World War II–era hero (reintroduced around when I got out of college) named Doctor Mid-Nite. He was a physician who could see in near darkness, striking fear into the hearts of villains and his patients who opted for the budget colonoscopy. Well, not only can you have super–night vision yourself, that power also comes with big, black eyeballs, like Jean Grey's when she goes all Phoenix. (If you're not an X-Men fan, maybe you shouldn't be reading this book.)

All it takes is putting drops of chlorin E6 (a chlorophyll derivative) directly onto your eyeballs, which causes your pupils to swell up and take over. It also allows the light receptors in your eyes to perceive very dim light in various colors. The research comes from studies of the deep-sea dragonfish, which would make a badass name for the superhero you are trying to emulate: Behold, the mighty low-light vision of Dragonfish!

Recently, a couple of biohackers (which seems to be a new term for jackasses) tried the drops out, scooping Dr. Ilyas Washington, professor of ophthalmology at Colombia, who had done trials on mice since 2007. Upside? You can creep around in the dark and see everything. Downside? It leaves you with a splitting headache, and if someone flips on the lights, you can permanently damage your vision.

ICEMAN

Many comic book characters, from Mr. Freeze to Iceman to Frozone, can withstand extreme cold, and they all owe a debt of gratitude to Eugene O'Neill.

But every year, the Garth Monks, whose name makes them sound like a hybrid country music/Gregorian chant group, take to the mountains of Tibet, strip down, soak some sheets in water, and hang out meditating. In minus-30-degree caves.

Their method is pure mind control: They have trained themselves to imagine that their cores are holding a furnace and their bloodstreams are like pipes, distributing heat to their extremities—so much heat, in fact, that their bodies cause the water in the sheets to evaporate, and they have to soak them all over again. It *is* awesome, but it *does* sound like a monastery full of dudes trying to cover up their bedwetting problem.

 One monk traveled to England in 2004 and demonstrated this amazing power for a BBC camera crew, concentrating on heating up his hand. In a matter of minutes, he raised the surface temperature of his skin by 22 degrees. I know what you're saying—no, wait, I don't know what you're saying, because that would mean I have a superpower of my own.

Don't confuse them with Wim Hof, who is impervious to cold and once climbed Mt. Everest in a tank top and shorts. Not only did he stand on top of the world dressed like Richard Simmons, but he guaranteed he'd never get a North Face endorsement. Who needs overpriced down jackets!?

UH ... THE WOLF FROM MIGHTY MOUSE?

Antonio Francis always makes the Internet's list of people with superpowers because he can cause his eyeballs to pop out of their sockets, then is able to pull them back in just by, I don't know, jacking up the blood pressure to his brain. At best, I would put him in the Legion of Substitute Heroes, since his mutant power puts him closer to a squeezy stress toy than anything particularly useful. While his power is definitely greater than that of the lame Wonder Twins, I mean, come on.

In any event, he makes my list because I actually met the guy when I was working on *The Tonight Show*. We were doing a live audience bit that was essentially a midnight talent show, and we usually included a ringer. Evidently, Antonio draws his powers from fermented spirits, because by the time we taped his segment at five in the afternoon, he had consumed six of Milwaukee's finest energy drinks and could barely speak. I guess if you can make a room full of people recoil in horror, that would drive you to drink—and, in my book, make you a superhero.

HEIMDALL

Remember Heimdall in *Thor*? Idris Elba in a Vegas-showgirl hat, guarding the gates of Asgard? He's a superhero, right? Well, the toughest part of that job is not battling bad guys trying to get in, or doing isometrics to keep the Gun Show looking good, or even not getting invited anywhere because you're working 24/7. It's that you have to *stay awake all the time*. That is a superpower, one he shares with Thai Ngoc, a Vietnamese man who—after having what he assumed was the flu (but that I'm suspecting was radioactive NoDoz)—hasn't slept in 43 years. And this started before anyone was able to binge-watch Netflix. Doctors believe that similar to the phenomenon that allows dolphins to remain awake around the clock, parts of his brain take turns sleeping and never do so all at once. This enables him to stay awake without going utterly insane, which would happen to normal sleep-deprived folks in only a few days. There are other famous insomniacs, like Paul Kern, a Hungarian soldier who never slept after taking a gunshot to the head. But, to be fair, would you?

BATMAN

While many people prefer monks who make beer, I have to say my favorites are the monks of the Shaolin Temple, from the Hunan province in China. David Carradine played one on the old TV series *Kung Fu*, but that's where the similarities stop—Shaolin monks generally don't do the kind of stuff he did in his spare time. But they represent the gold standard of martial artists, and the Dark Knight iteration of Batman relies heavily on their techniques and weapons.

In the real world, when the monks need to raise money for the temple (which has been around for more than 1,500 years), one of the masters takes a group of the younger guys out to demonstrate their martial arts skills for paying audiences. Kind of like the Chinese Acrobats of Taiwan, if the acrobats could silently kill you without moving—which they might well be able to do and will be the subject of my next screenplay.

I was lucky enough to meet them when I was producing *MTV Beach House* in 1996. Right before my disbelieving eyes, one guy did a back dive (sailor's dive!) onto concrete with no hands, landing in pole position on the top of his bald head. Another monk caught a bunch of arrows that were shot at his face. When the cameras were off, we learned that the monk who had snapped a Louisville

Slugger with his forearm had also snapped his forearm, never indicating the slightest pain and even setting the bone himself.

Discipline makes these guys able to perform beyond human limits. If I had even one one-hundredth of their concentration, I would have remembered how I was going to finish this sentence when I started it.

Oh yeah, I was going to talk about the Fist of Monkey.

My son Dashiell was just six months old when we were shooting that segment, and I was lucky enough to have room for Johanna and the kid to stay with me while I worked at the Malibu Beach House. I was holding Dash, and the old Master from the Shaolin Temple came over and started talking to him. They conversed in gibberish (or so I assume) for about five minutes, and I was starting to get worried we were in a *Little Buddha* situation right there, but then, as if cued by the old man, Dash turned to me and snatched the glasses from my face. The Master was delighted, and clapped his hands. Then he brought his men over to show them. The translator gave us the play by play: "The child has the Fist of Monkey. Whatever he wants, his body does. There is no society or civility or manners trained into the baby, and it makes him powerful. To use your body to its limits, you must un-learn those conventions that make you think before you act, and just *act*." So my son has the Fist of Monkey going for him. Which is nice.

Chapter 8

OUR PLACE IN THE FOODCHAIN

I once got to work with John Milius for a few years. Milius was one of the screenwriters for *Dirty Harry* and *Apocalypse Now* and cowrote and directed *Conan the Barbarian* and *Red Dawn*. The project—a biopic about Hawaiian surfing pioneer Duke Kahanamoku—never materialized, but the experience stays with me. Working with Milius was like pulling a log up to the fire, listening to someone else's stories, and figuring out what you could contribute to the conversation. One day, John was preparing to give a speech honoring his school chum Steven Spielberg, for whom he had written the final draft of *Jaws*. In it, he mentions how he took Spielberg surfing in Hawaii after that movie, but the director of *Always* wouldn't go in the water with him. "I just made sharks the biggest villain of all time," Spielberg explained. "They have it in for me."

Of course they do. They have it in for everybody. They're *sharks*. We exist below sharks on the food chain, and that gives us a primal fear of them. And

I mean a real primal fear, not like that movie *Primal Fear* with Richard Gere. I mean, Laura Linney was a revelation, but you can see Edward Norton's actual teeth marks in the scenery.

Kevin Smith has often discussed the reason *Jaws* has terrified him over the years: He doesn't want to go out of this life as some other creature's poop—or even give up a part of himself that way. And I admit, it would be very distressing if I were to lose a leg to a great white, to have to live with the idea that my leg—one of the two that, as a writer, I really only occasionally use—had been chewed, swallowed, and turned into shark turds and was now being re-eaten, maybe for the tenth time, by some slithering, bottom-dwelling creature in the ocean's icy depths.

By the way, I'm not talking about things that will just kill you, like everything in Australia. I'm talking about things that kill you and then go on to *eat* you.

Sharks aren't the only things above us in the food chain. There are plenty of other things out there that will make your sphincters chew gum in the dead of night. Here is a smattering of things that will chew you up and shit you out—I shit you not.

BEARS

Bears are cute. They do cute things like rub their backs on trees and help raise little boys in Kipling novels. They are mascots for sports teams and appear on the California state flag.

But bears are also *terrifying*. Depending on the type of bear and time of year, they can be two and a half to five times stronger than humans, pound for pound. That's orangutan strong, and brown bears can weigh up to 1,500 pounds. And bears are scavengers, so if you have any food in your camp, or in your car, or in your pocket, they will come and take it. A grizzly bear can flip over a 2-ton dumpster looking for food. They're like raccoons from the planet Krypton.

Brown bears are more dangerous than black bears, but neither variety usually bothers people unless food is scarce. Most guidebooks advise you to hike in groups when you're in bear country, sing songs, talk loudly, and break sticks. But really, all that will get you is the ability to say you went out singing, talking loudly, and breaking sticks after you're dead.

Polar bears are ruthless, and they make it look easy. Now that the polar ice caps are melting, they routinely go into settlements and try to eat people. You can google a famous *National Geographic* video of a photographer in a titanium shark cage, waiting out on the ice for a polar bear to come look. The cage, designed to withstand a 1,000-pound shark attack, completely falls to bits when the polar bear simply smacks it.

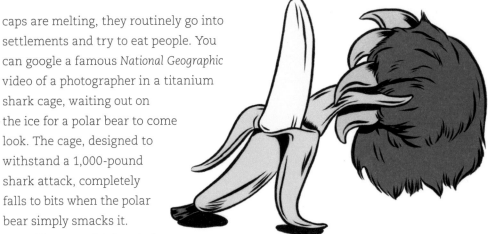

And beware of Asian sloth bears: They're killers because they have poor eyesight and can't run very fast, so they lash out at anything that even gets near them.

On a recent trip to Yosemite, my horrified family got to see a car that had been broken into by a bear. According to the guide, bears used to have trouble with the door handles and broken glass, but now they've figured out how to get in cleanly: They simply hook their claws into the rubber seal between the door and the roof and peel the door back like it's a banana. Then they rip everything, even the steering wheel and dashboard, out of the car.

So, don't stop for a bear and let him nose around your car. They know how to get in.

And if their sheer size, strength, teeth, and claws aren't enough, there is another way for bears to kill you: when you eat their meat. In Colorado, the government had to outlaw serving bear at exotic-game restaurants because, due to the hibernation process, bears tend to have a lot of dangerous parasites in their flesh that will eat you from the inside out.

So, hide your pic-a-nic baskets!

LIONS

More than 700 people a year are killed by lions. In Tanzania, the most common way to get killed by a lion is by using an outhouse, or otherwise defecating outdoors. (In case you were wondering, using this book is an inadequate defense against an attacking lion.)

Lions are scavengers and opportunists, though we have exalted them, used them as symbols of royalty and movie studios, and, essentially, lionized them. From time immemorial, until recently, humankind assumed that lions did all the killing and hyenas stole their food; however, it's the other way around: Hyenas—which are actually more closely related to cats than dogs—kill 95 percent of their own food. And when lions do hunt, they usually let the lionesses do it.

If you think you can face down a lion (which weigh up to 420 pounds for a male and 280 for a female), try picking up a 5-pound housecat that doesn't want to be picked up. You will lose some blood, some dignity, and the right to shop in that cozy little bookshop ever again. Now, multiply that by however much you have to multiply it by to get to 420 pounds, and you get the idea: You're the mouse.

If you want to watch a young Melanie Griffith get attacked by a lion, just stream the 1981 movie *Roar*. Tippi Hedren (Melanie's real-life mom) and Noel Marshall, her husband at the time, founded a wild-animal rescue center and decided after ten years to film a movie where their characters get attacked by predatory creatures, including lions. And to make it more realistic, *they really got attacked*. I can only assume that the birds in *The Birds* just weren't scary enough for Tippi, who broke her leg, while Marshall got gangrene and cinematographer Jan de Bont—who would go on to direct *Speed*—got scalped by a lion. And little Melanie got mauled, requiring facial reconstructive surgery, years before the typical Hollywood actress gets it.

It is possible to get lions to work for you. In southern Kenya, Dorobo hunters let the lions kill a wildebeest, then stroll in, acting tough, and haul ass with the meat. (That's one way to get the evening gnus.) However, they're still *lions*—if the hunters aren't quick about it, the lions will attack and eat *them*.

GATORS AND CROCS

Crocodilians (crocs, gators, caimans, etc.) are all related, and they are all horrible.

Alligator was coined by Shakespeare because he needed the rhyme in *Romeo and Juliet*. (It was a bastardization of the Spanish *el ligardo*, meaning "the lizard." Bastard!)

Crocodiles and alligators are in different families—and have killed more people than the Five Families. Gators have more rounded snouts, and crocs' are pointier, but it's not like you're going to have to tell them apart in the wild unless you are in south Florida, in which case you have a whole other series of problems I will not go into here. (Things get weird in south Florida, is all I'm sayin'.)

These things are all brain stems and teeth, and when you see one in real life, you get a visceral reaction to run. But it won't do you any good, because they sprint about twice as fast as a human—and that's a human that's running from an alligator!

When Kevin and I were shooting a lot of field pieces for *The Tonight Show*, we went to Gatorland in Orlando. Even though the average gator grows to be only about 15 feet long, they had an old-timer there that was closer to 20 feet long. When we went into his habitat—which contained about thirty gators—our guide told me not to worry too much about getting attacked: They had just fed it a 25-pound frozen turkey. And even if the gators decide to attack, they would take the slowest man down, eating him while the rest of us escaped. We all looked at Kevin—those Vans he wore weren't going to get him too far.

But that part of the shoot went fine, and thankfully, Kevin was able to go on to make *Jersey Girl*. But the last thing we did at Gatorland was the Gator Jumparoo, which was truly terrifying—really the stuff of nightmares. They had a system of covered docks that wended through a mangrove swamp, with pulleys on wires crisscrossing in certain areas. The idea here was to take a 5-pound frozen chicken, clip it to the wire, slide it out over the water, and watch as ferocious alligators dove, then leaped straight out of the water, sometimes as high as 8 feet in the air, and swallowed the chicken whole. The first time it happened, we all spit a little dirt in our pants.

Crocodilians kill more than 1,000 people a year around the world. The killing action is to bite with the big jaws, then use their powerful tail to spin, tearing their prey apart. A 20-foot Nile croc can leap out of the water and rip the head off a wildebeest so fast that they are back in the water before their prey's body has had a chance to fall over (which is really the best way to eat raw wildebeest, in my opinion).

An addendum to this story comes from the first shoot week of our low-budget horror/morality play *Killroy Was Here*, in Sarasota, Florida. It being a horror movie, we were shooting overnights for six straight days, and our big stunt night took place in Benderson Park, an enormous, man-made competitive-rowing arena. Knowing how Kevin feels about alligators, I sent a team of production assistants the length of the park to cover up any sign that warned people there were alligators about (PAs are expendable, or at least more expendable than having our director unwilling to come out of the trailer). Everything was going great, we had a terrific crew and cast and were having fun . . . until the woman who runs the entire park dropped by and said, "Whew! Real monster just over there. You see him?" I tried to play it off that WE had a real monster (Killroy), and he was right over there, but Kevin knew what was what. After a brief discussion, Kevin agreed to keep working if we stationed PAs around the location. I never *did* get a head count . . .

FIRE ANTS

It's true that mosquitos are the deadliest creatures to humans, thanks to how they spread malaria, yellow fever, West Nile virus, and a host of other diseases. And, yes, they do "eat" people, if you consider that tiny drop of blood to be eating. But now we're talking about animals that kill you for food. And the fire ant, horrifyingly, qualifies.

They come from South America, but these miserable little assholes have made their way to the southern United States, Australia, and Southeast Asia. They are just better at being ants than other ants: They breed faster and spread faster,

and rather than being prone to getting stepped on, they will kill livestock and other animals with their unique system.

Other ants bite, then spray acid on the bite, which is terrible. But these guys bite you only so they can get a good grip, then use a stinger in their abdomens to inject an alkaloid venom that can cause anaphylactic shock. At the very least, you get big sores like paintball hits that more often than not turn into pus-filled lumps. Worst of all, there are usually a number of them crawling on you: When you notice one or try to shoo it off, it releases an alarm pheromone and *they all bite at once!* Assholes!

I speak from experience here. When Johanna and I got married, our friends had a party for us on their ranch in south Texas. It was beautiful—there was a swimming hole, a margarita machine, and . . . fire ants. I got stung, and I pulled up my pants leg to reveal about a dozen bites that took a couple of weeks to go away. But when my father looked at his own legs, there were hundreds of bites. He needed antihistamines for days. Welcome to Texas!

Fire ants are the kind that build bridges with their bodies to cross small chasms, often sacrificing the ants at the bottom of a structure that can span up to 10 feet across. But if the whole colony can't make it, they grow new queens and drones with wings, and they just fly to the next field. As of 2015, invasive fire ants cost the US economy more than six billion dollars a year in destroyed crops and livestock and the expense of pesticides.

THE MOST DANGEROUS GAME

That's a reference to a movie based on a short story by Richard Connell, wherein a master hunter on a tropical island targets the only animals that can reason— humans. That story was the proto-*Predator*, in that once the hunter bagged his catch, he didn't eat the poor guy, but made a trophy of his head (though *Predator* went one better by keeping Carl Weathers's spine, too).

But there are people who eat people. Let's start with practicing vampires. Ordinarily, they consume only the blood of others. There are clubs wherein people exchange their own blood for recreational-beverage purposes—and there are other, worse clubs. I know this only anecdotally, since (and I'm not naming names) we had a guy at the old *Tonight Show* who was always asked to stay home whenever there was a guest escorted by the Secret Service. I asked him why, and he surprised me with a full explanation of his lifestyle choices, which mostly involved doing vampire stuff. You would think that a creature of the night would not have such a big mouth, but I guess all the better to put them fake fangs in.

But the real eaters of human flesh are the cannibals. It is a practice so fearful, so deeply ingrained as taboo in our brains, that in the extremely rare event that civilized folks have had to resort to it, they become famous *for all time.* Donner, party of six? Make that three.

There is no detail grislier than reading about the butcher with the Donner Party who cut up the dead, then labeled the packages so survivors wouldn't have to eat a loved one. Or the Uruguayan rugby team whose plane crashed in the Andes in the 1970s and who, after their rescue, detailed how difficult it is to eat a foot. All the little bones, you see.

And then there are cultures that practice cannibalism. Captain Cook, upon overstaying his welcome on what he named the Sandwich Islands because why not, was captured and killed by the native king at what is now known as Cook's Bay, after which his remains were shared among the other island kings. We now call those islands Hawaii, because "the Sandwich Islands" is stupid. (I think they got tired of the jokes: "Ooh, I'm the hero of Sandwich Island!" Or, "How did you get to Sandwich Island? By submarine?")

Setting aside deranged serial killers and that guy in Miami who thought he was a zombie, there are still cultures out there that practice cannibalism. Fiji used to be called Cannibal Island—another bad name—but Christian missionaries ended that custom in the mid-nineteenth century (ironically, with the body of Christ).

But the Korowai tribe of Indonesian New Guinea allegedly still has a culture of cannibalism. Rather than believing in germs, they think people take sick due to an evil spirit, which they then eat after their pal dies. (And you thought Irish wakes were crazy.) They also eat outsiders, and the last famous person to fall victim to this practice was an heir to the Rockefeller fortune, who disappeared in New Guinea in 1961. But the natives felt sick afterward—he was just too rich.

Chapter 9

REALLY GROSS ANATOMY

Of all the gross things you could come across in this filthy world, the grossest thing of all would have to be the human body. One school of thought chalks it up to evolution: We naturally fear, recoil from, or are otherwise repulsed by human bodies, alive or dead, since contact with other people is the best way to make us sick. This is balanced with our need for contact, unless you're Howie Mandel.

Taken as a whole, the human body is pretty disgusting, and all of your efforts to make it otherwise just prove the point. For example, you may want to slather those hands in Purell whenever possible to keep the gross things there from spreading all over your gross eyeballs and into your filthy mouth. But all that seems to do is kill off one kind of organism that is eating your dead skin and crapping all over it, only to be replaced by a superbug that does the same thing but also makes you sick as hell.

But even if you isolate parts of the human anatomy for consideration, you will find they are pretty goddamned gross. And if you eat a McDonald's cheeseburger, you can be one hundred percent positive that you have something gross in your body. But even if you don't, rest assured that—based on the following list and beyond—your body is a wonderland of mind-bending repulsiveness. Especially if you're a guy. Enjoy!

MITES IN YOUR EYELASHES

I wanted to start out small, with tiny creatures that are not quite parasitic but still eat parts of you—and make their pees and poops in your eyelashes. In the follicles at the base of your eyelashes, to be specific. Every time you blink—upward of 28,000 times a day, by some estimates—a gallery of little critters has a front-row seat to all that cornea squeegeeing. Sometimes, twenty-five or more mites *per eyelash follicle* feed on the oils on your eyelids. The older you get, the more likely you are to have them, with more than 80 percent of folks over the age of fifty likely to have their own herd of eyelid oil–guzzling grossness.

Don't look up a picture of them on the Internet; I did, and now I need to go to a hypnotist to have that memory taken out. Otherwise, even though they are not visible to the naked eye, whenever I close my eyes I swear I can see them, because they are so bizarrely close to the naked eye. Even though they are harmless and technically not a disease, there's a cure: Wash your eyelids with baby shampoo, getting into the corners and nooks and crannies and really giving them the thrice-over. Or you could use tea tree oil (or soap), which is guaranteed to get rid of the pests but might make you do a full Rod Steiger from *The Amityville Horror.*

BACTERIA

All this talk about gut bacteria being good for you seems to be there to distract us from the fact that we carry around a *lot* of bacteria with us in our horrible bodies. Scientists estimate that there are as many bacteria cells on and in us as there are actual human cells. Since bacteria are small, however, they weigh between just 1 and 3 percent of our body weight. But if you weigh 200 pounds, you might have 6 *pounds* of bacteria. (That number goes down if you take Metamucil regularly—pun intended—so I'll let you guess where most of this mass lies.)

Your armpits are also a great place for bacteria to grow, and while a small percent of BO is caused by hormones, most of that smell comes from bacteria eating your body oils and pooping where they live. Still, "That's not me, that's bacteria poop," is not a great excuse for permanently funking up someone's car. Setting aside beneficial gut bacteria, there are other bacteria that might make someone rich someday: For years, beauty care companies have tried to harvest the bacteria in the crook of your arm, which makes your skin there soft and supple. Your next face cream might just be bacteria based.

DEAD STUFF

According to several sources, roughly 300 million cells out of the 10 trillion to 50 trillion cells in your body die every minute. At least 30 million of those dead cells are red blood cells, meaning that that same number of red blood cells is manufactured by the bone marrow every minute (and they live for about 120 days). Then the spleen, which also stores blood, filters out the old red blood cells and breaks them down, sending them to the liver and the gall bladder, then into the lower gastrointestinal tract by way of the duodenum. This is just one reason you don't want to keep your gall bladder if they take it out. This is also why your poop is brown (not red, due to the process)—if live red blood cells make it into the colon, they turn your poop black. (This is the "grossest parts of the human body" section, if you recall.)

You lose about 5,000 to 8,000 cells from the largest organ you have—your skin—every minute. Dead skin cells account for most of the indoor dust you breathe (which qualifies it as a gross thing in your body), and it's why old libraries are so dusty—the leather of book covers is cow skin. Where does all that skin go? Well, a lot goes into your bed, which is why a two-year-old pillow can be as much as 10 percent dust mites, dead dust mites, and dust mite poop. How does Howie Mandel sleep at night?

THE MOUTH

You don't have to be in the middle of the Human Centipede for your mouth to be gross. Sarah Silverman isn't the only one with a dirty mouth: Even if you brush, floss, and Listerine the shit out of your mouth, you're not actually getting it all. The cleanest mouths have upward of 100,000 bacteria on *each tooth*. Only one in ten people use a tongue scraper, so at any time you might have 3 tablespoons of bacteria colonizing the back of your tongue and throat. And even though they cause bad breath, many of these little organisms react to invading bugs by proliferating and choking the environment out, keeping you safe from infection.

Since mucus is made from blood plasma, it's a great growing medium, which is why having postnasal drip is like giving a cascading supply of food to your disgusting, disgusting throat. And here's a good reason to floss right now: The same molecule deposited as plaque on teeth is in arterial plaque, and your gums allow stuff to get into your bloodstream and contribute to heart problems. That's how chewing tobacco gets absorbed and gives you a nicotine buzz. And if you are chewing tobacco, that is definitely the grossest thing in your body right now.

POOP

Setting aside the fact that poop is funny to talk about—even phrases like "duty roster" crack me up, and the fact that most comedians perform with a stool they never sit on must mean something—it is an important part of the way your body works. Poop is full of stuff you don't want: live and dead bacteria, fat, salt, insoluble fibers, dead cell husks, and, anecdotally, complete corn kernels that must have somehow reformed themselves, because you are positive that you chewed them up. Some of the living bacteria in your stool is there because it likes to eat that other stuff (except the corn), producing waste that includes methane gas. So when you fart, you are farting one big fart of millions of microfarts executed by little farting bacteria cells that, if they found farts funny, would have great cores from laughing all day.

Food and liquids and saliva basically turn into the sloshing gunk from which you get nutrition, and after the cilia in your intestines gets what it can, some of the water is absorbed from the now-waste material, making feces. The longer you hold it in, the harder it gets—so, you know, when nature calls, pick up the phone. (And since we're on the subject, coprophilia is a condition in which someone likes to eat poop. According to Parul Agarwal, MD, assistant professor of gastroenterology and hepatology at the University of Wisconsin, where my son goes, eating your own poop is safe if you're not sick. But eating someone else's poop definitely is not. So stick with the Pittsburgh Plate Job and leave the Cleveland Steamer to the pros.)

Chapter 10

STUPID SCIENTIFIC STUDIES

The scientific method of inquiry is amazing. Essentially, you take a look at some behaviors, data, cause-and-effect relationships, and so on, then you formulate a theory to explain the phenomena. A scientific theory is different from how most people define a theory, since it is based on reliable data and in most cases can be proven by experimentation. When you think *scientist*, you think *experiments* and *studies*—whereby, say, the Professor is able to prove why Gilligan has super-strength.

However, each year we find that a lot of money is being spent on studies that ask the stupid questions—and often get the obvious answers. Add to that SCIgen, a program that automatically writes fake scientific studies *that actually get published*, and you'll find a lot of dumb studies out there. Despite my love of real science—and my trust that it not only makes our lives better but also, in many ways, makes us better people—my favorite studies are those that ultimately prove that even stupid people can be scientists. And often, they're paid for by your tax dollars!

WHY DO YOU THINK THEY CALL IT "DOPE"?

During an average nightly news broadcast, viewers can hear more than two minutes' worth of drug side effects read by pleasant announcers during the commercial breaks. One reason there are so many dangerous side effects in our drug supply is that the approval process for drug trials is much shorter than it used to be. And new side effects of existing drugs, especially recreational drugs, are being discovered all the time.

In 2014, researchers discovered that rats prefer classical music to jazz, but when you inject them with cocaine, they become jazz lovers (and they just don't stop talking about it). A 2015 study found that the drunker you are, the more likely you will participate in unsafe sex—provided you are not too drunk to have sex. A study from Duke University in 2012 claims that smoking marijuana makes you stupid; an early 2016 study of twins, as the *Washington Post* reports, proves that smoking marijuana does not make you stupid; and at the end of 2016, a twenty-year study published in the journal *Addiction* proves unequivocally that smoking marijuana does indeed make you stupid. (Maybe studying the effects of marijuana makes you stupid?)

But it's not all bad news. A variety of studies show that drinking beer can have various health benefits: It increases bone density in men, suppresses inflammation, improves heart health, prevents kidney stones, aids digestion, prevents Alzheimer's and cancer, and makes you skinny. I will let you pick which of those studies you think are stupid.

FOOD

The average nutritionist will tell you that a diet high in plants and low in things with faces and mothers is going to improve your health and make you live longer—though eating like a vegan just makes your life seem like it's taking longer. But a great nutritionist will tell you that we really don't know dick, that eating butter doesn't seem to bother some people and that Chinese people have been eating right for the last thousand years, even though their average life span is shorter than that of Americans.

In 2012, a study seemed to show that dieting makes you fat because it changes your brain and hormones. But a 2016 study proves that healthy dieting helps you lose weight and live longer.

A 2005 study of diet sodas claimed that they were better for you because you weren't eating sugar, and they had no side effects, but in 2016 a new study showed that the combination of preservatives and artificial caramel coloring suppresses the production of enzymes in your gut and makes you fat. A UCLA study in 2017 claimed that high-fructose corn syrup suppresses synaptic function and makes you stupid, so nobody really wins in the cola wars.

In 2016, a team of psychologists from Plymouth University claimed to have proof that you could curtail cravings for sex, alcohol, and food of all types simply by playing *Tetris* for three minutes (just hearing the song works for me). UC Santa Barbara is the proud employer of two fluid physicists who proved scientifically that it's easy to spill your coffee when you walk with it.

And a U of Saskatchewan professor has published a paper that claims that because boogers have a sweet, sugary taste, they should be eaten, and that the reason little kids do it is to strengthen their immune system. I personally prefer snot with a hint of umami.

RELATIONSHIPS

So, you think that relationships are the most complex, unique, and daunting things to understand on the face of the earth, and that above love and sex, tolerance is the key? Obviously, you don't read *Modern Psychology*, because for the past 200 years, researchers have been sciencing the shit out of relationships.

The best ones look at the differences between men and women. Agreeing on a comfortable room temperature is a source of strife in couples, but University of Utah researchers have cleared that up by showing that women are almost always colder than men, because their hands are more slender and tend to get colder. And when it comes to meeting someone, we are pretty clueless: Only one in three men can accurately detect when someone is flirting with them, and only one in five women pick up on men's moves. Sad, but not as sad as the results of a 2014 study that claims that in familiar relationships, men have orgasms 22.2 percent more often than women, and lesbian women reach orgasm 13.1 times more often than heterosexual women. (Unless the man is having sex with the lesbian—that never works.)

Short men make more money than their wives and are 32 percent less likely to divorce them. My theory? The men are too tired from going out there every day to bring home the jack, and who wants to have an affair with an exhausted short guy? Mixed-race men and women are more attractive to potential partners, though only men are attracted to friendly strangers. And when things go south, men are 41.6 percent more likely to develop a drinking problem in response to an unsatisfying relationship.

HEALTH

It seems like much of health reporting is based on finding a bunch of really old people who are still walking around and having sex, and trying to figure out what it is about their lives that makes them so healthy. In 2016, the Italian town of Acciaroli on the Amalfi coast got a lot of attention because a UC San Diego doctor went to the beach there and noticed everybody was in their nineties and older, smoked cigarettes, and seemed healthy. Not only does this prove that this doctor is incredibly bad at picking fun beaches to go to, it also proves that eating lots of rosemary (the herb) and sardines (the fish) "smooths out" the arteries, cuts down on inflammation, and keeps you sexually active.

At least you feel like you're learning something there. Toronto researchers published a study in *Annals of Internal Medicine* that concluded that obesity is bad for you. The Harvard School of Public Health determined that coffee can make you feel upbeat. U of Florida researchers found that if a person likes to ski, going skiing makes them happy.

And God forbid if you want to start diagnosing your infant. Neurologists at Emory University found that babies that made less eye contact were more likely to be autistic. Then again, some babies that made eye contact—even above-average amounts of eye contact—could have autism, too. But don't fret: You can ameliorate some of the social symptoms of autism by letting the kid play video games, specifically goal-oriented role-playing games. The competitive need to understand someone else's methods while simultaneously seeing the world through another character's eyes is supposed to train spectrum patients in empathy.

My favorite source for this kind of improbable research into the fundamentally obvious comes from a website with a fundamentally obvious name: *Improbable Research*. Not only do they report on the annual Ig Nobel Prizes, they have a wonderful podcast and detail lots of stupid studies. A recent highlight comes from a fluid-dynamics research team at Virginia Tech. They were investigating just how people—after swimming, for example—were able to remove the water from their ear canals. You know, since they're too small to fit fingers and most towels. Using data obtained from the CT scan of a human head, they printed an ear canal on a 3-D printer, put water in it, and investigated several ways to get the water out. The answer? Shake your head. Preferably to one side, with the target ear pointing downward. (Not even a) Problem solved!

ANIMALS

I'm not going to get into the argument that using animals for any kind of research is pointless. This section is dedicated to showing other ways that researching with animals can be astonishingly inane. A researcher at University of London's Royal Veterinary College issued a report detailing why King Kong could never have existed. Turns out he's too big. Unfortunately, Peter Jackson's *King Kong* existed, complete with Naomi Watts's cliffside vaudeville dancing and that weird platonic love scene on Central Park Lake.

You can believe your eyes when you see a pig appear to be happy while wallowing in mud: A review of sixty-six research papers in a 2011 issue of *Applied Animal Behavior Science* supports that perception. A 2014 study found that dogs do learn to associate words with objects, but they do it differently than humans do. (I guess they do it doggy-style.) British researchers determined that adults talk to their dogs much in the same way they do to infants, although they give the dogs more orders and ask the babies more questions. In a 1993 study of elderly dog walkers, a three-man team came to the conclusion that people with dogs tend to meet and talk to more people than those without dogs, and—surprise—they talk more about dogs.

And, ignoring the fact that cats are just below humans in the list of species most responsible for other animal extinctions (which we get some credit for), *Smithsonian* magazine reports that while the research is sketchy, cats are smart and don't care about their owners as much as dogs do. Maybe they're smart enough not to care about these utterly useless scientific studies.

Chapter 11

SIDE EFFECTS MAY INCLUDE...

Having worked for Jay Leno all those years, I really got addicted to the news. Even the nightly news, which I can only assume puts me in the same category as people with irritable bowel syndrome, atrial fibrillation, migraines, non-24 syndrome (nothing to do with Kiefer Sutherland), and men with erectile dysfunction who like to bathe with their wives in separate, outdoor bathtubs. Because ten out of the seventeen ads on the *NBC Nightly News with Lester Holt* (the Lester Holt the better) were for prescription drugs. And a full three and a half minutes of the broadcast, within those ads, was devoted to an earnest voiceover artist pleasantly reading out the horrible, horrible side effects you risk encountering if you take them.

I'll start with the bad stuff. And I have to admit, even though I know it's coming for most of these pills, it still astonishes me to hear "May cause suicidal thoughts." And that is for the stop-smoking drug Chantix—and the acne drug

Accutane, and the allergy drug Singulair, and the flu drug Tamiflu, and the cholesterol reducer Lipitor and . . . the list goes on and on. I guess it's a profitable side effect, since if you are having suicidal thoughts, you might take the whole bottle. Cha-ching!

And what are the worst ones in that category? Antidepressants. Not just a few—it's turning out to be all of them. It's like putting a warning on a bottle of Excedrin that says, "May cause headaches."

There's a diabetes drug with the warning "Do not give to children under 8. Children 8 to 18 should not take it, as it can cause developmental issues." Wouldn't it be easier to say, "You must be over 18 to take it"?

Look, if I'm going to take something that causes liver and kidney failure, heart arrhythmia, and in some cases death, it better not be Chantix. It better be something where it makes me feel better than I have in my entire life, gives me superpowers, does what all the other drugs do combined, and lets me see into the future. I guess that shows you just how addictive cigarettes are if people are willing to risk all seven plagues of Egypt just to stop smoking.

I should note that since the Food and Drug Administration started getting the drug companies to pay for their own drug trials in the 1990s, Big Pharma has flipped it on the watchdogs, shortening the trial period to as little as two years. So, many of these side effects don't come out with the drug: They are taken from anecdotal evidence of patients who were given legal prescriptions by their doctors in real-world cases. We are now just biological beta-testers for new and exciting drugs.

What follows is the briefest of looks at side effects throughout the years. But, just like the ads all say, if you're interested in taking a prescription drug—and in some cases, not even prescription!—consult your doctor. And that phrase was repeated ten times in a half hour on TV last night.

WEIRD FOOD INTERACTIONS

Grapefruit. Famous for having its own silverware. Centerpiece of a 1980s fad diet. Essential ingredient in a Sea Breeze, a Greyhound, and a Salty Dog. And, if combined with certain prescription drugs—which one-third of adults over the age of fifty-five take—can result in *death*.

How can the most popular citrus fruit without a county or a song named after it send you on the big Dirt Nap? Grapefruits contain the compound curanocoumarin. For those of you who don't know what curanocoumarin is, first up, read a book, and second up, it blocks the enzyme cytochrome P450 3A4. What's cytochrome P450 3A4? Well, you've made it this far, so I guess I'll tell you. It's used to metabolize a ton of drugs, up to eighty-five of them currently on the market. Without this enzyme, the drugs can reach toxic levels in the bloodstream, and fast. Statins like Lipitor are on the list, but there are lots of other popular drugs on the list, like the blood thinner Plavix. Welcome to the new millennium! That grapefruit might kill you!

So now you're nervous. You can't sleep, so you decide to have a soothing cup of chamomile tea. But if you're taking an anticoagulant like wayfarin, you might wake up dead! That's right, just like a karate man that bruises on the inside, the mild blood-thinning property of chamomile tea can combine with your prescription blood thinner and cause internal bleeding. And you thought it was just for wussies!

So, no chamomile tea, but you do want to be healthy. You have a salad with chickpeas, kale, lettuce, and broccoli, even though broccoli tastes like crunchy garbage. But wait—you've just dosed yourself with foods high in vitamin K, and now your wayfarin doesn't work! Quick! Get some chamomile tea!

Now you're starting to get depressed. A world without chamomile tea and broccoli? Your doctor, his prescription pad ever at the ready, decides you should take a nice daily dose of the antidepressant Nardil, which is an MAOI (monoamine oxidase inhibitor), but that doesn't matter, now that you're even keeled again. Life is looking up. You've taken a break from cooking, so all you have at home are some ripe bananas, some dried sausage, beer, a little spinach, and a nice blue cheese. You throw the spinach away because it looks too much like kale and you're afraid of it, but that doesn't matter, because all these foods

cause you to have a big spike in your blood pressure. You get severe chest pain and clamminess and start to sweat. This is ten times worse than when you found out that Stilton cheese has shit in it (which could be right now—sorry to spring it on you like that).

So, when you start taking drugs—any drugs, particularly those you have to take every day for the rest of your life—read the fine print. All of it. Just because you're not mixing things with alcohol doesn't mean anything these days.

BOTANICAL DIETARY SUPPLEMENTS

Time to learn the difference between homeopathic remedies and herbal supplements. Homeopathic remedies are nonpharmaceutical, nonreactive, and essentially noneffective remedies invented by Samuel Hahnemann in 1896 with the idea that if a substance causes the symptoms of a disease in a healthy person, it will cure that same disease in a sick one. Now, if that makes sense to you, you might also get the thinking that diluting the substance in water or alcohol makes it more potent. Personally, I don't.

Further, the dilutions are epic. It's called the C scale: 1C is a dilution of 1 part substance, 100 parts water. And it's logarithmic, so 2C is 1 part to 10,000. 6C—the most potent—is 1/1,000,000,000,000, or 1 part per trillion. Most labs can't even detect that for most compounds.

So homeopathy = pffft.

But you can get in some real trouble with herbal supplements, even though people think they're completely safe because they're "natural." Lots of hard drugs fall into this category because they are derived from botanicals: Morphine, penicillin, and aspirin are technically herbal supplements, and all can be toxic. Even milder botanicals can be dangerous when combined with prescription and over-the-counter drugs.

And St. John's wort, a natural remedy for depression, can interact with a broad spectrum of drugs. St. John must've been the patron saint of side effects, because if you're also taking an SSRI (selective serotonin reuptake inhibitor) like Celexa or Prozac, it can radically increase the many side effects of the SSRI and give you what's called serotonin syndrome, a rogue's gallery of symptoms including tremors, temperatures of up to 106 degrees, and diarrhea. St. John's wort can increase the effects of many sedatives, decrease the effects of many blood pressure and blood-thinner medications, stop your HIV drugs from working, and, if you're taking a birth control pill, allow you to get pregnant. Even dextromethorphan, the *DM* in Robitussin DM, can lead to serotonin syndrome when taken in combination with St. John's wort.

And that's just one "natural" danger lurking in your local Whole Foods Market.

Your vitamin have CoQ10? It can negate your blood thinner, and you might throw a clot.

Drinking a Cape Codder? The cranberry juice has the opposite effect, making blood thinner work overtime, and can cause bruising or bleeding.

Echinacea is a popular method for treating the common cold, just as crossing one's fingers is a popular method for increasing the odds of winning the lottery. But it can inhibit the breakdown of caffeine in your bloodstream and make you jumpy.

Kava can help you sleep, but combined with booze it can cause liver toxicity, and if you're detoxing from an opioid with Butrans or its relatives, it can put you in a coma.

Even ginger, ginseng, and garlic can mess with blood thinners and cause problems.

This is just the tip of the iceberg; many, many herbal supplements have active ingredients and potential interactions. Even though you can buy it in the magazine stand at the airport, melatonin can cause you some real problems, like daytime drowsiness and depression. Actually, just being in an airport can produce those same side effects.

ZOLPIDEM (AMBIEN)

Yeah, you're all familiar with this one.

More and more studies have come out showing that the best thing you can do for your health is get a full eight or nine hours of sleep a night. Some of us have trouble with that: The older we get, the more we have to wake up to pee. You might live in a city and be a light sleeper. Or, if you're like me, when you have to get up to an alarm, you wake up throughout the night to repeatedly make sure you haven't slept through it.

So, it's super-important, and you would do anything for a good night's sleep. And Ambien is the one to take you out. But is it worth it? This is a drug that the FDA had to add an additional warning label to more than two years after it had been in general use—and that label included this language: "After taking AMBIEN, you may get up out of bed while not being fully awake and do an activity that you do not know you are doing. The next morning, you may not remember that you did anything during the night."

Admittedly, that could describe how some people felt after the last election. But while that sounds somewhat innocuous, the list that's inside the bottle, on that insanely folded-up piece of paper with fine print that makes your credit card privacy statement look like a Dr. Seuss book, is super-specific. Activities might include "sleep driving," which is *driving while asleep*. It's common enough that a New Jersey DWI lawyer advertises he'll get you off with the "Ambien defense." (Get you acquitted, that is.)

Another activity is making and eating food—and, from some news reports, that can mean eating an entire lasagna while standing in front of the open refrigerator. And talking on the phone. To someone else. While asleep. Also, walking around (which is astonishingly vague—"Yeah, it's four in the morning, and I'm just walking around the neighborhood in my pajamas with no shoes"). And, oh, having sex. Like, you are fast asleep, and by all reports not even dreaming, when this pill you took decides to chuck your partner a hump.

And, in some cases, not your partner.

Taken together, these side effects seem to be more like demonic possession than anything else. You take an Ambien, your soul leaves your body, and someone else slips in there, has a quickie with the wife, and eats lasagna.

I think if I ever took an Ambien, it probably wouldn't work. I'd be up all night worrying about what I'd do while I was asleep.

ROPINIROLE (REQUIP)

Look, Parkinson's disease can be devastating, and I applaud any effort to try to ameliorate the symptoms and let people get on with their daily activities. So along comes Ropinirole, also known as Requip, which can possibly cause constipation, loss of appetite, dizziness, increased sweating, nausea, vomiting, and general weakness. And in case being a dizzy, vomiting sweatball doesn't get in the way of your daily activities, it also can make you randomly fall asleep without warning. Like when you're driving a car. And in this case, you will probably end up crashing; it's not like sleep driving with Ambien, where the worst you do is pick up a box of Ho-Hos and bang the neighbor.

So, yeah, it's a lawsuit-y drug. And it may not even be necessary for many Parkinson's patients, according to the American Parkinson's Foundation, which states that simply increasing the amount of exercise you get every day can help with the symptoms. And while to many of us, getting off the couch may be a horrible thought, this thing gets worse. Like, lose-yourself-ruin-your-life-and-end-up-on-the-street worse.

First, a little background. People with Parkinson's gradually stop producing dopamine, the neurotransmitter that helps regulate movement, emotions, and the feeling of pleasure. This drug is supposed to mimic dopamine's functions. But there's a difference between a job well done and overdoing it, because in some cases, users of Requip have experienced hypersexuality, compulsive shopping, gambling addiction, and other abnormal behavior. Now I'm talking like the clinicians that write that pleasant side effects ad copy. No. This is bad. Let me put it in terms you'll understand.

A woman in Minnesota took Requip for a few months and, even though she was a devoted mother and wife, started going out and sleeping with strangers. As much as she could. She began to deplete the family savings by compulsively buying things at stores and online, ultimately ruining their fortunes by gambling away all their money and accruing huge gambling debts. (Seems like when you lose your inhibitions, the first thing to go is that you care that you lost your inhibitions.) It all happened too fast for anyone to help.

These fundamental personality changes may seem to be rare—I mean, if a lot of women were going full-on Salem-witch-trial batshit looney, we'd all know about it and this drug would be pulled, right?

Nope. This is a Google page-fiver. Meaning, if you google the name of the drug or keywords about the story, you'll have to page down—a lot—before you get to reports like this. Keep going a little farther, and you might run across the Mayo Clinic study that shows, in a trial of 720 patients, *one in five showed one or more of these drastic symptoms*—meaning if the Spice Girls all started taking Requip, one of them would change her name to Fucky Spice.

Oh, and the kicker? It's also prescribed for restless leg syndrome. I guess if you're going to take this drug, you might be a bit of a gambler already.

WHILE IT WOULD BE EASY TO JUST MAKE THIS SECTION ABOUT SCIENCE FICTION, THE FI SECTION ALSO INCLUDES THE SCIENCE *OF FICTION.*

In other words, we look at things that folks may have believed but have been proven false, like urban legends. Or we look at fictional characters and, using science, try to discover a deeper meaning, like diagnosing the *Fat Albert* gang to understand what the hell is wrong with those guys. We also look at ideas that are patently false and tell you why they're false, like so-called evidence of the supernatural.

But we're not about bursting bubbles around here. Just because people like to use the supernatural as evidence of an afterlife—and time and again,

supernatural occurrences have turned out to be hoaxes, mass hysteria, or confusion on the part of the observer—that doesn't mean we're claiming that there is no afterlife (even though it is perfectly obvious that there is no afterlife). It is not up to us to tell anyone that mystical, cross-dimensional energies don't enable your spirit to travel into new realms of consciousness; that is for my next book, *Shit That Is Obviously Shit, and Why You Shouldn't Believe It.*

Fiction is actually incredibly important to science. So many concepts and technologies from *Star Trek*, for example, have inspired generations to change our world: Cell phones. Hypodermic medical spray. Tablet computers. Having sex with people in costumes. And this is just a sampling.

And it works the other way. In college, I took a course called The History of Science in Literature. It was half History of Science, and half How Science Influenced Literature. Popular articles on how electricity could make a dead frog's muscles twitch inspired Mary Shelley's *Frankenstein,* for example. For my part, I did my senior paper on how the discoveries of the atomic age inspired the upsidasium cycle on *Bullwinkle.* (I knew myself even then.)

So, get ready for fake things people thought were real, real things that arose from fake things, fake things that have always been fake to the point that we shouldn't have to prove them fake anymore but for some reason we still do, and fake things that came from real things.

And wherever I find myself in a real spot, I'll just fake it.

Chapter 12

CRYPTIDS

So many common phrases come from seafaring travelers through the millennia. "By and large" on a sailing ship refers to two kinds of sails being set. All sails aloft is called "the whole shooting match." "Touch and go" refers to a dangerous landing where a boat cannot tie up, but can only touch land before departing. And so on.

But for the hundreds of phrases and words sailors brought back to land, there are many more mythological creatures that their groggy (i.e., had too much grog) brains invented during long, strange sea voyages. Since most sailors in the early 1800s could not swim (!), pretty much anything below the surface was fair game for wonderment and exaggeration. General illiteracy did not mean they were ignorant, but it did allow their stories to survive purely by word of mouth. Play a game of Whisper Down the Lane with four other people, and you'll get some wild stories yourself.

And then there were the hoaxes. The more disconnected the stories were from motives and profit, the more outlandish (and believable) they were to the general population. Like Mulder said, *we want to believe*, and that's spawned cryptozoologists—pseudoscientists on the hunt for cryptids, creatures whose existence is unsubstantiated yet hasn't been completely disproved.

We touched on a couple of these in the Urban Legends section earlier on, in chapter 2. It's not stupid or foolish to believe this stuff—it's just human nature. The same human nature that brought us Bigfoot and White House press briefings.

MERMAIDS

I know what some of you are thinking: There has to be some truth to this whole mermaid thing. Disney is not in the business of promoting weird, magical creatures—*unless you count almost every movie they've ever made*. But old Walt would be spinning in his perfectly-normal-underground-temperature final resting place if he knew how mermaids probably came about.

From time immemorial, men have gone to sea in ships. And for the most part, women have not. So, after a few months of boners and nowhere to put them (pegboys notwithstanding), pretty much anything with a hole and a heartbeat starts to look pretty damned good to them. The accepted theory is that—by only making brief appearances at the surface in shallows that would founder a ship—sea cows were idealized as voluptuous denizens of the deep that would entice men to their doom. Now, take a look at a sea cow: Which half is supposed to be fish, and which half is woman?

After months at sea, I don't think even that mattered.

On a trip to Seattle with Kevin Smith, we visited Ye Olde Curiosity Shop, a place with lots of jars full of repulsive things like two-headed rabbits. The proprietors are in possession of what, for many years, was believed to be proof positive that mermaids existed: the full skeletal remains of a creature that appeared to have the body of a small woman and the tail of a fish. However, by most accounts this hoax is really the upper body of a monkey sewn to the lower body of a large fish, probably a tuna, with the skin mummified onto both parts. They call this monster "the Thing," and there are other "Things" at other roadside attractions. In my opinion, the real monsters are the dudes who spent an afternoon sewing dead monkeys and tunas together.

We got in a bit of trouble for the piece about the shop that aired on *The Tonight Show*: Kevin was struck by the incredible resemblance of "the Thing" to Joan Rivers (although her skin was tighter). That wasn't so bad, but when he compared the smaller, even uglier companion mermaid to Melissa Rivers, that's when the phones started ringing.

BIGFOOT/YETI

Oh, man, I'm going to lose some friends over this one. You can make fun of religion, you can make fun of the president, but when it comes to claiming that a large, hairy ape-man, living either in the woods of the Pacific Northwest (Bigfoot—the brown one) or the Himalayas (Yeti—the white one), you are delivering a metaphorical nut shot to some *true* believers.

Rick Dyer is a Bigfoot hoaxer living in Atlanta. While most folks who claim they've seen the Tacoma Yeti live in the Pacific Northwest, he does it from the humid comfort of the Jewel of the South. Just by faking a couple of photos and making a website, he has convinced many people that Bigfoot exists. And when people find out that for one brief, shining moment, they were allowed to believe, and then reality rudely crashed in on their *Harry and the Hendersons* fantasy, they *despise* this guy for it.

Growing up, we had *The Six Million Dollar Man*, a show that proved that Lee Majors could not act and gave The CW's *The Flash* the idea that, in order to show someone running really fast, it's best to just show them running in slow motion for some reason. On that 1970s masterpiece, Colonel Steve Austin, who

could lift an armored car with his bionic arm and bionic legs without somehow crushing his human torso, befriended Bigfoot—I mean, I say they're friends, but they probably didn't do a lot of hanging out—and introduced the world to the term *sasquatch*, which, in Halkomelem, a Native American tongue, means "large, hairy man."

If you're ever in Ochopee, Florida, first of all, I'm sorry, and second, you simply must visit the Skunk Ape Research Headquarters. Many a mullet has passed through those hallowed doors. And there is so much to parse in that name, so let's get to work.

First, the Skunk Ape is the Everglades version of the Sasquatch. There is no Native American name for it because the Native Americans, like the rest of the world, know that "skunk ape" is a stupid name. Second, evidently, "research" is conducted here, which seems to take the form of making fake artifacts and quoting drunks' and idiots' firsthand accounts. And, finally, *headquarters* conjures the image of the nerve center of a thriving system of research outposts keeping daily tabs on the daily activities of the culture of skunk apes living there. Of which there are none.

As for Yetis? Forget the Yetis. There are no Yetis. Or Sasquatches, or, even more obviously, skunk apes. Jesus Christ, why don't you read a book now and then?

MONGOLIAN DEATH WORM

I am including this one because it sounds like some kind of exotic ninja move used by sex assassins and in all likelihood is the inspiration for the sandworms in Frank Herbert's *Dune*.

But in actuality (or nonactuality—it is a cryptid, after all), this is a fantastic beast that has been around in Eastern folklore for thousands of years. It was first detailed for westerners in Roy Chapman Andrew's book *On the Trail of Ancient Man* in 1926, wherein he stated that, on asking a gathering of Mongolian officials about it, "none of those present had ever seen the creature, but they all firmly believed in its existence." Bingo! *Cryptid*.

This thing is gross. Take its Mongolian name, for example: *Olgoi-khorkhoi*, or "large intestine worm." Supposedly, it swims beneath the surface of the sand, creating waves that allow it to be detected. Or maybe the wind is just blowing.

It can attack from a distance, either by delivering an electric shock or spraying a venom in your face like some nightmarish result of a disembodied colonic. It seems to like the color yellow, so if you're in the Gobi Desert, leave your Gorton's Fisherman costume at home. And here's the most unbelievable part: The natives claim that touching it, even in the slightest way, causes instant death *and horrible, racking pain!*

Which you can't feel, because you're instantly dead. So they gotta work on that part of the story.

And thus the tail wags the dog: In 1990, a team led by Ivan Mackerle went into the Gobi in June or July, which is when the Mongolian death worm is most active (which is actually never, so we're dividing by zero here). Using cues gleaned from *Dune*, these knuckleheads actually built thumper devices to draw the worms to the surface. Even more unbelievable? They went back in 1992.

Since then, the worm has been the focus of TV shows like *Destination Truth* and even *Beast Hunter* from the National Geographic Channel, which just goes to show how far National Geographic has slipped since *Explorer.*

LOCH NESS MONSTER

I will try to make this fast, because there's some interesting stuff here, even though we all know it's a hoax.

First up, the monster was never described until after the brontosaurus (which was gone from science for a while but is back—see chapter 1) was discovered in 1877. Since then, people around the Scottish lake have been telling tales of a long-necked creature with a menacing demeanor who wasn't Sarah Silverman telling you to put your money in a credit union (I love Sarah Silverman, but c'mon, she does have a freakishly long neck), ambling across the road or rearing its fearsome, prehistoric head out there in the water.

Proof positive seemed to come in 1934, when a London gynecologist photographed an unusual phenomenon, breaking through the surface with a bulbous head and a long, erect, shaftlike neck. It was later learned that, not only was this a fake serpent model, it was constructed on the back of a toy submarine—which was not filled with seamen but with wind-up toy clockworks that enabled the device to plow through the wetness. There is more to the story, of course, but I have run out of innuendo to tell it.

Further "proof" came in 1952 when John Cobb, an English racing motorist, was killed during a boating speed trial on Loch Ness. The official report was that an errant wake caused him to lose control of the boat, but we all know what *really* happened: An errant wake caused him to lose control of the boat.

Still, Nessie fans were convinced he was killed by the monster, and not by an untested jet boat going 200 miles per hour. What kind of fool would tempt fate on Loch Ness? What kind of fool would tempt fate in an untested jet boat? The answer to both questions is John Cobb.

By 1993, when the Discovery Channel's documentary *Loch Ness Discovered* first aired, nearly every shred of "proof" of the Loch Ness Monster had been debunked. Even the gynecologist's photo was analyzed, and showed the blurry shape to be a model only about 2 feet tall.

So, if you have an untested jet boat you want to take for a spin, Loch Ness is as good a place as any. And if something goes horribly awry, you have only yourself to blame.

THE KRAMPUS/BLACK PETER

By now, you might have picked up that I have no patience for cryptids. But this one is particularly interesting, in that it is a Scandinavian tradition that essentially describes what we think of as Satan. And it's gaining in popularity around the world, again probably due to the Satan connection.

Another reason I'm interested is that I have close friends who have crossed paths with the Krampus (pronounced "krump-US"). But first a description.

Up Scandinavia way (and other parts of Europe), where the cold winds blow, Santa (Sinter Klaus) doesn't go it alone. Sure, he shows up with a bag full of toys for the good girls and boys, but alongside him is his little pal, the Krampus.

Here's a dude who also has a sack, but it's not for dropping off—it's for picking up. The Krampus is essentially a satyr, half giant goat, half man, with cloven hooves and horns and a pointy tail and—whoops! Here comes Satan!

So, while the good kids get Scandinavian Erector sets and Ikea furniture, the bad kids get *kidnapped by the Krampus and thrown in the sack!* Then they are either taken to the woods to live forever with the Krampus, *or they are carted off to hell!* Suddenly, the Grinch doesn't seem so bad, am I right?

It all happens on December 5, which is Saint Nicholas Eve, which explains the twofer. And as you may have guessed, the Krampus itself is a cautionary tale for those children who have pissed off their parents in one way or another. And it happened to a friend of mine, because, as though telling the tale wasn't bad enough, his parents actually dressed up and scared the shit out of their kids.

My pal Michael is a talented chef and father of three who grew up in the Swiss Alps. Dairy farms, alpine horns, yodeling, the whole bit. He and his little brother, ages six and five, were very good boys, but occasionally they might complain about having six hours of chores a day, or having to get up at four in the morning to milk the cows, or having to *grind the fucking salt for dinner.*

So, to teach them a lesson, their parents had a visiting uncle dress up as the Krampus—bearskin pants, leather jerkin, and terrifying mask, all of which the uncle had brought with him for just such an emergency.

At bedtime, which was like two in the afternoon, he burst into the house and chased the kids around the house, bag open, moaning that he was going to take the children away forever. The parents stood aside and were like, "Well, guess that's what happens!" The boys made it to their room, the uncle left, and neither of the brothers slept until just a few weeks ago.

Public Krampus celebrations are another thing entirely. This is a more Germanic tradition that takes place in festivals like Krampusnacht in Munich. When my beloved niece Martita was spending a semester abroad in Austria, she and her boyfriend had no idea what was going on when gangs of marauding assholes, dressed like my friend's uncle, were running around on a drunken rampage with clubs, riding crops, and switches. A couple of them whipped my niece in the legs, cutting through her jeans and opening gashes in her legs.

Fearing for her safety, her boyfriend pushed through a crowd of what were essentially extras from *A Clockwork Orange* and managed to get the two of them onto a bus. But as the bus pulled away, he shined on the dressed-up drunks outside with a "Nyah-nyah—we got away!" look. The gang blocked the bus, pulled him off, and punched his teeth down his throat.

And that rowdy, angry mob tradition is celebrated in more and more cities across the United States every year. Happy Krampus, everybody!

Chapter 13

BRANCHES of SCIENCE WE PRUNED -OR- SHOULD

If you have been paying attention, that's amazing, because I am mesmerized by the design they put together for this book, and I'm still writing it. And, if you've been reading sequentially, instead of just opening the book to any convenient page and reading until you need the Charmin, you'll have a pretty good sense of what the scientific method entails. You are way ahead of a lot of people, including so-called scientists of the past.

For those squat-and-peruse types, here's a recap: Analyze natural phenomena. Come up with a theory supported by those facts. Test that theory through experimentation or by collecting more data. If the glove doesn't fit, chuck the theory out.

Confirmation bias make this hard. You want to be right. You're going to go looking for data that supports your theory. You want those eggheads that never listened to you to be obligated to name something after you. Unfortunately, they don't. And even more unfortunately, sometimes your dumb idea spills over into popular culture, we tag an -ology onto the end of your cockamamie idea, and, *although your idea has been rejected by the scientific mainstream*, people faithfully follow it.

I'm happy to say that there is room for faith and science in your brain. But having faith in science itself? No. Just . . . no. Faith is the belief that something exists without proof. Science is not that!

Look at the 21-grams theory. No, it's not about six mah-jongg games played by women in their sixties, it's about the weight of the soul. In 1907, Dr. Duncan MacDougall of Haverhill, Massachusetts, was convinced that the soul was a physical thing, and therefore had weight. So, he put a dying man on a scale, made him as comfortable as possible, and waited for him to give up the ghost. The man did, and lost weight at the moment of death.

Then MacDougall tried the same thing with a bunch of dogs, and—you guessed it—no weight loss. Since dogs don't have souls, that could mean only one thing. To an idiot. Who was looking to be right.

He performed the same experiment on six more people and got different results each time. Two people gained weight (does this soul make me look fat?), so he threw those results out. The other results were all over the map. Add to that the fact that it was impossible to determine the precise moment of death, the scale contraption was unreliable, and one patient gained *and* lost weight after death, and you know this theory is on shaky ground. Perhaps it actually was on shaky ground, which also would account for the crappy results.

But, since everyone wants this kind of proof in their lives, it was published in the *New York Times*, and Bob's your uncle. (What this alludes to is my theory that everyone in the world has an uncle named Bob. I just have to find the right data set to support my theory.)

PREFORMATIONISM

To hear English physicist Dr. Brian Cox say it, many physicists are drawn to physics based on the idea that it is an intellectual pursuit that obeys certain rules. Attempting to unify the disparate theories of physics is a wonderful way of understanding them, and it pushes you to finding how new rules might work.

That, and biologists aren't *really* scientists, which comes off as a joke but I think he's serious about it.

That's because biology is full of so many bizarre things. And, more and more, biologists have to understand other scientific disciplines in order to understand our world. Take, as an example, photosynthesis. How does the plant use the chlorophyll molecule in perfectly timed sequence with acquiring a photon to make water and carbon dioxide form organic molecules? *Quantum physics.* Pfft!

Well, there was a time when biology was so simple, a child could understand it. In fact, a lot of it seemed like a child had come up with it. A sexist, fundamentalist child who could tell you how to figure out the ratio of the sides of a right triangle but who didn't know crap when it came to where babies come from. I'm looking at you, Pythagoras.

His theory—which became known as preformationism—was that all the essential characteristics of offspring come from the father, and the mother contributes only the growth medium and the tank in which the baby grows. Furthermore, this applied to all creatures, which were all created at once—"Let there be light," and so forth. This lasted all the way to the seventeenth century by way of Aristotle and later Europeans like Galen, Colombo, and Fabrici. Jerks!

In 1651, William Harvey published a seminal (!) work stating that all animals come from eggs. So far, so good. But he went on to say that within the eggs, the creature in question was a super-miniature version of what the creature would become. And I mean, like, there was a tiny dude in every sperm, posed in cannonball position, just waiting to take the plunge in the genetic pool and grow up. Google the drawings: They look like turnips trying to cover up their dicks.

In one of science's greatest "What are you doing in there, honey?" moments, microscope refiner Antonie van Leeuwenhoek jerked off onto a slide and took a look at a sperm cell under magnification for the first time. No Mini-van Leeuwenhoek. Progress! Then he reasoned that because sperm is a complex structure, every single human sperm has a soul. Dang it, van Leeuwenhoek! I thought we were getting somewhere!

By the eighteenth century, there was a jumble of ideas as scientists tried to figure out why so much life and so many souls would be wasted in a money shot. It wasn't until Caspar Friedrich Wolff came along and argued for objectivity—meaning, take religion out of the equation—in scientific investigation that this became a moot point. Don't get me wrong—he still had his head very far up his ass when it came to miniature people in sperm—but, you know, baby steps.

Finally, just before the turn of the twentieth century, Wilhelm Roux and Hans Driesch experimented on developing sea urchin embryos and ruled in favor of epigenesis—the idea that organisms develop over time in stages from different contributing material. Meaning, they got it right finally—at least until we discover that they had their heads up their asses, too.

ASTROLOGY

For the sake of keeping the peace in a house where people believe that swishing coconut oil in your mouth does something (I forget what—I'm not on Facebook enough to get roped into those nutty trends), I will say that certain things about astrology have not been *disproven*. The same can be said for O.J. and the Trump administration, but there it is.

And just because Plato and Aristotle believed it doesn't mean jack. They also believed that if you act badly, you're going to feel bad. See above reference to O.J. and the Trump administration.

Distilled to its bare minimum, astrology encompasses the belief that your entire life is influenced in obvious and subtle ways based on when you were born, where the planets were at the time, which part of the night sky was in a particular spot over the horizon, and how the moon played into all of this. Further, your days are influenced by what is up with said celestial bodies now. Where does predeterminism play into all of this? Why don't employers read your charts to figure out whether you're the right guy or gal to hire? How come Scorpio is a water sign? (I mean I get Cancer, 'cause it's a crab (see page 15), and Pisces being a fish, but scorpions?)

I should note here that the General Social Survey, a nonprofit organization out of the University of Chicago, reports that from 2005 to 2015, the number of Americans who think astrology is "not at all scientific" fell from 65 percent to 53 percent. So, nearly half of Americans think astrology has some science to it!

There's a group called the Magi Society that purports to find a direct correlation between reliable data sets (stock market, marriages, etc.) and astrology. But they do it in such a bombastic and condescending way that they must all be a bunch of Virgos.

Personally, I'd like to know whether there is a correlation between when single people started asking, "What's your sign?" and the divorce rate.

The fact that astrology is ancient does not mean it's true. "A million ancient Babylonians can't be wrong!" is just a stupid statement. Come on, people, the idea here is that we are supposed to be getting *less* superstitious and *more* scientific about things. How much astrology went into the making of the iPhone? Probably none. And if you have an astrology app on your phone, you are a newly minted kind of douchebag that has never been seen before. (Fact: Most astrology sites and apps are in part phishing schemes to get your birthday.)

I have to admit, however, that astrology did play a part in the birth of my daughter. Her older brother had to be delivered by C-section due to the combined enormity of his parents' skulls. So, when it was time for Daisy to come into the world, the doctor decided to schedule it. There we were, one early March morning (you're not getting her birthday, phishing site!), heading off to Cedars-Sinai while my sister-in-law watched our son Dash, ready to meet our new child. Johanna was extremely ready to have this baby. And just as we were leaving for the hospital, the phone rang: We had gotten bumped.

Another expectant mother, scheduled for the afternoon, wanted her daughter to have a morning sign. She insisted. The doctor was embarrassed, but what, you're going to fight with a pregnant lady?

And for years, at Joey's Gym, preschool, elementary school, wherever, if we met someone with a daughter who shared a birthday with Daisy, I would muster all my powers of mental telepathy to read whether it was *that* mother, and, if so, destroy her life with my brain.

Because astrology is bullshit!

REFLEXOLOGY

If you've watched *Pulp Fiction*, you will know that there has long been a debate as to whether foot massages are a simple favor or whether they're foreplay. Foot fetishists notwithstanding, I believe that any intimacy beyond fleeting eye contact is somewhat out of bounds.

But there is another side to the foot-massage discussion: the dubious science of reflexology. See, you know it's a science because the word ends in *ology*. And you know it's dubious because it's advertised on the subway alongside the chiropractors and Dr. Tattoff, the tattoo-removal guy. Another clue to the dubious nature of reflexology is that it is supposed to cure pain and diseases in the other parts of the body through manipulation of the feet (or hands—which is more or less intimate?).

Here's what happens: We find ourselves in a "doctor's" office with a bunch of Chinese charts on the wall and some very distressing stains on the waiting-room carpet. Is that incense burning as aromatherapy, or to cover up whatever made that stain? The doctor explains to you that the pain in your chest and left arm are due to energy fields called qi (or chi) that run throughout your body. Just like your body, those qi have to stand on two feet, so they extend from all parts of the body to the feet and hands.

The idea here is that the reflexologist starts working his or her thumbs into your feet like a teenager texting *Dancing with the Stars* votes. A little bit of this, a little bit of that, check the chart, more working of the thumbs–and, voilà! No need to visit the cardiologist—that pain in your chest was just misaligned qi! You feel better, maybe because energy fields need realignment, but probably because you just got a foot massage from a stranger—which, let's face it, is foreplay.

Want to find a study that proves reflexology is false? Just google "scientific basis for reflexology." Then google "gullibility." "Magical thinking." Then sit back and see how many bullshit ads Google starts serving you while you peruse those articles.

DOOMSDAY PREDICTIONS

Science and natural philosophy have, for millennia, gone hand-in-hand with making predictions about when God is going to take his ball and go home, and everything goes "boof." This is a tragic branch of science—not just in the subject matter, but in with what people do with it. Because on the one hand, Armageddon is pretty terrifying, and on the other hand you have *rational authorities* saying it's on the way.

So, people act on it. If you ever find yourself saying, "Let's start a religion!" a doomsday cult is a smart way to go. There are so many ways to pick a date and support it with so-called scientific evidence. Plus, your followers aren't going to exactly be all that attached to their possessions, so the hat usually comes back full after you pass it. And the easiest cult recruits are folks who find life pointless: At least with Doomsday coming, they're right about one thing! Maybe find a cave, throw in some Nikes, and it's a party.

Sir Isaac Newton was a pretty smart guy. Came up with the calculus. There was that gravity thing. Head of the Royal Society. But his obsession, the thing that drove him to a life of anxiety, late nights, and confirmed bachelorhood, was that he believed that the Bible contains semi-mathematical clues that constitute a hidden prediction for the end of the world. I mean, why not, because it's not like the Bible isn't explicit about detailing crazy, horrifying things.

After more than forty years of work, Newton was relieved that it wasn't coming until 2060. But he's not the only scientist who predicted the end of the world. If you visit a present-day scientific gathering, awards ceremony, or panel discussion, it's bound to come up, because *most scientists believe the world will end.*

Yes, some of them are referring to our sun burning out in billions of years. Others think we're due for a mass extinction event based on something smashing into our planet, or Earth's molten core slowing down and providing less magnetic protection from cosmic rays, or climate change destroying the delicate balance that has kept us going for so long.

Oh, and there's the Doomsday Clock. It was created by the people behind the *Bulletin of the Atomic Scientists*, a scientific journal that raised the alarm in 1947 that nuclear weapons would spell the end of the world. They set it at seven minutes to midnight, and it inches closer and backs away based on international news of nuclear proliferation or other signs of potential global catastrophe. Right now, it's at two and a half minutes to midnight, the second closest it's ever been, and the scientists are pinning it all on Donald Trump.

Fake news! Sad!

UFOLOGY

Yep, it ends in *ology*—and shares an *o* with UFO, which makes for good branding.

Here is a branch of pseudoscience that, if it actually comes up with some results, will have to change its very name to *ifology*, because the unidentified flying objects will be identified—presumably as the spacecraft of aliens from distant planets who have decided to identify themselves to us now that we can orbit the Earth a few miles up and crap in a plastic bag.

Occam's razor is the name of the philosophical principle that, given the choice between two possible explanations for a phenomenon, the simpler of the two is more likely to be right. However, in pseudoscience, things work differently.

When an unexplained phenomenon occurs, *there must be an explanation*. Some people—most people, in fact—would rather ignore the simple explanation and go for the *most interesting* explanation.

What's that? You live in an abandoned speakeasy in the desert north of Cabezon, California? And, mere months after nearby Edwards Air Force Base starts testing top-secret aircraft, missiles, and other payload-delivery systems requiring the frequent deployment of high-altitude weather balloons, you start seeing weird lights in the sky? And your old boss, Howard Hughes, fired you because *he* thought *you* were crazy?

Which are you going to believe—that there's a secret government program you're not in on, or that you have uncovered a world-shaking secret *only you know?* Well, if you're George Van Tassel, you're going with door number two. And, just to cement the deal, you will come to believe that you were contacted by said space aliens and instructed to build the Integratron, a three-car-garage-sized Van de Graaf generator that uses static electricity to shock the age out and make you young again.

Sounds more like he was visited by Jim Beam and Jack Daniels.

I took comedian Tom Green up to a ufology viewing field in Washington for *The Tonight Show*. It turned out that every single one of the dozen or so ufologists we met—who had all seen and made contact with space aliens—had suffered some form of traumatic brain injury earlier in life. One sweet lady had been in an accident with her son—both drowned, but only she was resuscitated. (Now she visits with him when he comes down on a space ship.) They all kept seeing what they insisted were spacecraft orbiting the planet, but every one of these sightings turned out to be satellites. (We brought along a night-vision camera.)

They all had something else in common: They all wanted to believe.

Chapter 14

ANYTHING TO THAT SCI-FI TROPE?

All entertainment is vaudeville. And the first rule of vaudeville is, "If it's entertaining, it's three times as entertaining if you do it three times."

Based on that, I must conclude that certain overused elements in science fiction could potentially be infinitely entertaining, because they have been used and reused, over and over again. This practice not only enables studio executives to relax their sphincters a little bit by staying on familiar ground, it also can lead to some of these tropes registering as true in our common-concept set.

Each story is not limited to a single sci-fi trope, either. Take *Alien*, for instance. I mean, the stunning visuals of uncharted planets promise to take us into the unknown and really blow our minds with concepts that would paralyze first-time pot smokers for ages to come. And yet, most of the conceptual innovations are in the art direction: The ship. The eggs. The creature. And look how thick the soles of those sneakers are! This must be the future!

But then the film resorts to many, many expertly timed, beautifully crafted, cleverly deployed horror-movie tropes. The characters get picked off one by one in classic horror/suspense style. The tense buildup leads to an explosive sound—of a cute cat that's not the monster. Don't be too funny, or you'll be the next one to get it in the neck. The last one to survive is a woman—who inexplicably strips before the final showdown. And so on.

Set aside movie tropes per se, and focus on the overused sci-fi concepts. The aliens are malevolent planet destroyers bent on using their strange adaptations to wipe us out. Evil takes the form of a robot that looks human but does not have a soul and is incapable of making a moral decision. And the true evil? A profit-hungry corporation that cares nothing for human life if there's money to be made from the discovery of an alien creature. And in the face of all this technology, tension, and terror, people are just people, no matter where you go.

I should probably point out the difference between hard science fiction and soft science fiction. While "hard SF" sounds like it refers to fanfic wherein Spock and Bones get it on, it actually pertains to sci-fi that adheres to a standard of scientific accuracy. Speculative science is acceptable, as long as it is based on sound theories. 2001: A Space Odyssey is an example of hard SF, meaning it follows the laws of physics and so forth. Soft SF is more character driven, or imagines speculative societies, often based on a high concept with little scientific merit. Children of Men, for example, shows the dystopian fallout of a world wherein all women have gone sterile, a soft-SF idea.

Tropes exist in both. Maybe the biggest one is that no matter how much everything changes, the hardest thing of all to change is human nature. Or, to quote Kevin Smith, "Even in space, people are assholes."

HUMANS GOOD, ALIENS BAD

This one bugs me the most because it is so damned hypocritical: A group of explorers encounters an alien race bent on destroying all of mankind. If you take a look at history, this is quite literally the opposite of almost every new encounter between societies.

I mean, sure, plenty of explorers were killed by the natives when they arrived. In fact, it almost happened to me: I was on my way to see my college girlfriend at Princeton when I got lost in the Jersey Pine Barrens and went to a *Deliverance*-style gas station. If I had been driving anything nicer than a VW Rabbit, had had more than twenty bucks on me, or had looked delicious, I would have disappeared right then and there, I swear to God.

But for the most part, we arrive, and everyone else dies. And we celebrate it. Amherst College's fight song alludes to the fact that Lord Jeffery Amherst was a pioneer in germ warfare, giving smallpox blankets to the natives during the French and Indian War:

> *Oh, Lord Jeffery Amherst was a soldier of the king*
> *And he came from across the sea,*
> *To the Frenchmen and the Indians he didn't do a thing*
> *In the wilds of this wild country*

What kind of a dick do you have to be to laugh over that one every time there's a game? The kind that goes to Amherst, it turns out.

What if an alien race were some kind of simpler creature, like an insectoid or Matthew Modine? Would we use our benevolent intellect to improve their lives? No. We would kill a few, dissect them, put a few more in zoos, and figure out if we could eat the rest.

However, in a universe of (in our minds) infinite possibilities, there's a good chance that the first alien race we encounter will be so advanced, so far ahead of us intellectually and technologically, that we will be the insects and the Matthew Modines in their eyes. Maybe we won't get intentionally wiped out by an alien race. Maybe we'll just get unintentionally wiped off their windshields.

HUMANS USE JUST 10 PERCENT OF THEIR BRAINS

Luc Besson loves him some sci-fi tropes. *The Fifth Element* is full of them: In the future, we will have flying cars, but they will look just like regular cars and be all beat up and stuff, people will wear outlandish clothing because we will have seen it all and be bored with the Gap, and people will be able to put up with Chris Tucker. But it wasn't until *Lucy* that Besson decided to base an entire film on a worn-out premise that has been disproved time and again.

You've heard it before: You are walking around with an almost unlimited potential that simply needs to be unlocked. No, this is not a Dianetics ad. (And who came up with *that* nonsense? A sci-fi writer!). I'm talking about your brain. You use just 10 percent of it, and the rest is wasted, like bagged salad from the grocery store. If there were only some way to use *all* of your brain, your life would be limitless. Or the movie *Limitless*. (I guess it was a TV show, too, but it got canceled, putting it up there with ironic titles like *The Neverending Story* and "I Will Survive.")

I get the fantasy behind it. Your brain is like Luke Skywalker stuck on some dumb planet harvesting moisture for a living. But that's just your outside self. Inside, your brain has the potential to save the universe.

Except that it doesn't. You use all of your brain. You don't use all of it all of the time, of course. You have to sleep. And all through your waking hours, you focus and use just part of it to drive your car, follow the news, have pangs of regret for things you said, and think about going to the gym. This is called concentration.

And that's just your conscious mind. Your brain is doing all kinds of unconscious shit so you don't have to, like making your heart beat and your lungs breathe, producing hormones, processing sensory input, and checking it against what you know and feel, and doing all the housekeeping that a supercomputer needs to keep going. Also, part of the regular functionality of the brain is to invoke chemical processes that enable you to concentrate on a level so complex that it's hard for the human brain to fathom. For example, researchers have just discovered the process that enables you to suppress memories you don't want to think about *without thinking about it*. My mind would be blown by this if some part of it didn't suppress the memory of that previous sentence.

In fact, your brain uses up to 25 percent of the calories you eat every day. So, of course you're using all of your brain. If you were using only 10 percent of it, and then you went all *Lucy* on everybody and multiplied that by ten, you'd have to chow down 20,000 calories a day just to think, which is like four Bloomin' Onions from Outback. That's horrible!

I guess *Lucy* sort of transmogrifies into a creature with no body by the end there—which, if you ask me, is a dumb thing to do with an actress as beautiful as Scarlett Johansson. And, come to think of it, *Lucy* is just *Luc* with a y attached, like an adjective. Do you think that Luc Besson was saying he's really the result if you let ScarJo use her entire brain?

DESTROY THE MONITOR, DESTROY THE COMPUTER

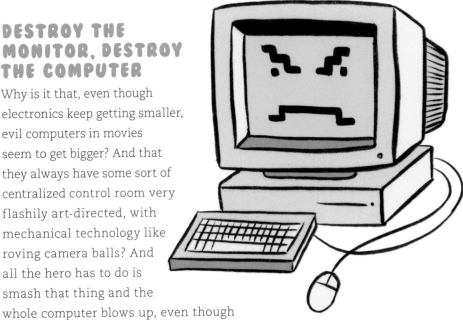

Why is it that, even though electronics keep getting smaller, evil computers in movies seem to get bigger? And that they always have some sort of centralized control room very flashily art-directed, with mechanical technology like roving camera balls? And all the hero has to do is smash that thing and the whole computer blows up, even though it's probably made of solid-state, nonflammable materials?

I, Robot, Eagle Eye, and non–Shia LaBeouf movies like Terminator, War Games, Superman III . . . the list goes on and on. Somehow, despite the idea that in each of these movies there's an evil computer that's taking over the world *by taking control of the network*, it still lives on only one anthropomorphized machine. And even now, there is no such thing as the goddamned *cloud*—it's a metaphor for out-of-sight servers that people can easily access over the Internet. No, it seems like the people making these movies are writing and producing them on a PCjr.

Come on! Even the special effects used to make these moves don't exist on one giant machine! When we made our humble little movie *Yoga Hosers*, I did the special effects: I composited Kevin Smith as the foot-tall Bratzi and supplied more than a hundred CGI Bratzis rendered on farms (big processing arrays of computers and graphics cards) in Canada, Germany, and the Philippines. It's a good thing that my machine didn't become sentient, because then it would all have moved to one computer and Shia LaBeouf could have come in and smashed it with a mop.

The really frustrating part of this for me is that computers already have taken over the world. We are no longer actively in our lives. Have to wait in line for more than one second? Whip out your phone and check Facebook. Driving your car? Why not text something like "U C John Oliver? LOL" and crash into oncoming traffic? Psychologists call it "passive locating," where your consciousness is more engaged with a continuity related to your online life, not your physical life.

The machines have won! I'm calling Shia LaBeouf and getting him to smash my phone right now! Shia, if you're reading this, grab your mop!

THE BUTTERFLY EFFECT

You know this one—someone is tasked with time traveling on some kind of mission, but they are warned: Do not fuck *anything* up, or you might irrevocably change the present! You could destroy mankind! You might give Chuck Berry the idea for rock 'n' roll, and your ugly siblings will disappear from that Polaroid you carry around! Then, when you get home, everyone will have pointy sideburns instead of sideburns that are trimmed with flat bottoms! It's a world gone mad!

This might have worked on *The Twilight Zone* back in the 1960s, when the technological miracle of watching moving pictures right in your home was still a thing. The only problem is, time doesn't work that way. Before we get into that, let's pretend it does and see how that works.

Say you go back to Nazi Germany to kill Hitler. (I would if I could, and in general I don't like killing people.) And say you succeed, plugging that bastard right above the cock duster with a Luger stolen from Goebbels. Everyone's like, "*Ach du lieber! You've kilt Mein Führer!*" And they lift you onto their shoulders and carry you out to the streets, and there's a spontaneous parade and even the Hitler Youth are like, "Fuck it" and join in the fun. (And then you infect everyone with all the modern diseases you're carrying that they have no immunity to, like you're Lord Jeffrey Amherst or something.)

Problem is, time—or space-time, to be more precise—doesn't work that way. Time is just one dimension of our multidimensional universe, and just because we perceive it sequentially doesn't mean it's plodding along linearly and we can jump around in it. The ability to perceive—or perhaps limitation of perceiving— time in a linear way is a survival trait that evolution has selected for. It's better for your consciousness to see the rhino charge, and then run like hell, for instance. Or to eat as a response to being hungry. Helps us survive, but doesn't give us insights into how that particular (fourth) dimension works.

You can't go into the past, because you didn't. End of story. And as for traveling into the future, you're doing it right now. If you want that to happen faster, travel at a high velocity, and time will elapse more slowly relative to you.

So, if you want to go dinosaur hunting, wait for science to bring the dinosaurs back. It's closer than you think. And we can all celebrate over a McMastadon burger.

EVERYTHING NEW TENDS TOWARD UNIFORMITY

In taking one's first steps into a cool, mind-expanding science fiction concept, it makes sense that you want to keep things simple: Focus on the concept; don't make the story too hard to tell by making it too complex.

Say you create a galaxy that existed long, long ago and far, far away. In it, there are thousands of planets that support life. And that life can be pretty fucking weird. You go to the cantina, and there's a dude with a butt on his face. You have a robot that's purpose-built to translate, and yet everyone understands all the weird languages anyway. There's an admiral that's really a fish in a onesie.

So, when it comes to the planets, even if there are wildly different creatures there, each planet has only one season going on, with an incredibly specific defining climate: The forest moon of Endor. The ice planet of Hoth. You don't hear Hothians saying, "I just adore this part of Hoth in the spring."

Forget planets—let's step it up and look at parallel universes. You're beaming up from a planet after some bullshit about getting more dilithium crystals— same-old, same-old—about to go back to your ship of very diverse, some-good and some-not-so-good folks, like you do. But there's a plot complication, and you get switched with your counterpart from a parallel dimension and end up in his universe, where *everybody* is evil! Except the people with the dilithium crystals, for some reason. And before you figure out how to get back to your own universe, you make some changes here, like getting your hot but homicidal alternative-universe girlfriend to realize that mean people suck.

So, you really haven't saved anybody or done anything except introduce the idea of subtlety to an entire dimension of beings.

I guess with global warming and the destruction of so many habitats, species, and ecosystems, maybe we are trending toward simplification, and the Morlocks and the Eloi aren't that far off.

Except there is this sort of overarching rule to the universe: entropy. The only real order is that things tend toward disorder. The only thing that's certain in science is that everything tends toward uncertainty. It'll all end in the eventual heat death of the universe, which is depressing—even Thomas Pynchon can't make it fun. The universe will end not with a bang, but with a whimpering, dull *bleh*.

So, if you're writing that sci-fi flick right now and you don't want to end up in my next book, get a little random, please. It will make the future seem so much more natural.

Chapter 15

THE SUPER-NATURAL

When Roy Horn (of Siegfried & Roy, the famous Las Vegas magicians) was mauled by their white tiger, something so obvious happened that people denied it. People just couldn't face it. Newspapers reported on it as an absurdity, and radio hosts started looking at it as though it were some kind of crazy conspiracy.

As Roy, God bless him (and God bless his fifteen costume changes with codpieces that gradually increased in size), lay in his bed, with parts of his body now in the digestive tract of an animal he considered to be a close friend, he started getting a very strange visitor. Strange not for his difference from the recovering Roy, but because he was so *similar*.

This guy looked exactly like Roy. But how could that be?

Let's step back for a second. In their act, Siegfried was the brains of the operation, the main event. It was he who started out in Austria, doing magic on the street. It was he who parted the curtain alone in the second act and picked the obvious shill from the audience. It was he who coined the most magical word in the cosmos, *Sarmoti*, for which I will never forgive him.

But Roy was the twist. It wasn't a partnership in the classic magician style: It was a magician and his assistant, but the assistant was a boy. Roy.

Roy, who got shut in one box and jumped out of another in an instant. Roy, who plunged into a pit of fire, only to gleefully shout "Haloooo!" from his rope swing overhead, his codpiece gleaming in the footlights. A crazy person might say there was some kind of trick involved—no one could be two places at once!

And now this? Another guy that looks exactly like Roy?

Theories abounded. It was a publicity stunt, people theorized, and Roy was actually unhurt and was caught leaving the hospital because he was going stir-crazy. Or Siegfried blamed Roy for the accident and was finding a new "Roy" to replace him. Or . . . or . . . *something*.

But the plain fact was that this was Roy's double coming to visit his injured friend, and—hold onto your hats—Roy had never really dematerialized in one box and rematerialized in another. The codpieces were real. The magic wasn't.

For all our rationality and scientific curiosity, part of the way we think is set aside for wonder. My friend David Regal (a writer on our show *Rocket Power*) is the world's foremost close-up magician, and I spent many wonderful hours at the office not working, watching him perform mind-bending sleights of hand and tricks of the mind. He demonstrated new tricks on me, and sometimes the explanation was more complicated than the trick. The simplest reason behind it is magic. But there is no magic.

David explains that you have to understand the psychology behind our belief in magic. There are times in our lives when we want nothing more than to ask, "How did they do that?" when we know exactly how they did that. The box was rigged. The legs were fake. *There were two Roys.*

The supernatural is magic. We believe in magic because we *want* to believe in magic. Not because there is magic. There isn't.

If there were, how in the hell would that tiger have been able to attack Roy?

The following "powers" are often seen in comic books. The X-Men's most powerful mutants do this shit. And, not coincidentally, many real people who are convinced they have these powers turn out to be schizophrenic or otherwise show signs of traumatic brain injuries. So, the next time Aunt Tessie sees a mourning dove, reads the signs, and tells you something bad's about to happen, do us all a favor and take her to a qualified clinical psychologist. And if something bad *does* happen, chalk it up to chance. (Or blame the Illuminati—those guys really do have it in for you, or so I dreamed while staring into my Nostradamus finger bowl!)

TELEPATHY

Here is how the standard telepathy test works: Two people sit across from each other. One of them looks at a card drawn from a Zenier deck—you might remember these cards from the opening scene from *Ghostbusters*, each with different shapes, wavy lines, etc. The person with the card concentrates with all their might, sending out the image of the card into the ether. Then the other person visualizes which shape the first person is thinking about. It's a one-in-five shot.

In a large sampling, you will find that no statistical evidence is discovered that supports the idea that people can think at each other like they're from *Beneath the Planet of the Apes*. But you *really* want it to be true, because it would be helpful in keeping future talking apes out of the Forbidden Zone. So, you take the statistical outliers that scored well, call it a talent, and say they're more sensitive than most. Bad science.

Now, it is true that people communicate, often in sophisticated messages, without talking. Couples who have been together a long time know what each other are thinking based on years of reinforced nonverbal cues. I, for example, know that my wife is annoyed with me because she is awake.

Don't be fooled by studies that legitimize telepathy. There are a lot of kook websites that like to bring up how Stanford has conducted many studies on telepathy and remote viewing for the CIA. Ultimately, they proved that telepathy and remote viewing won't be of any use to the CIA or anyone else, because these abilities don't exist, so the fact that the researchers took the CIA's money doesn't legitimize any of it.

This kind of communication between twins is often mistaken for telepathy. Twins might experience things similarly from a distance because they grew up together, spent all their time together, and are unconsciously routinized in the same ways. So, what might feel like random events tied together by twindom usually turns out to be coincidental based on these routines. And, again with the confirmation bias, the one time out of a thousand that one twin burns himself and the other one feels it leads believers to believe. If "feeling transfer" were real, however, it would make masturbating even more embarrassing than it is.

But science is coming to the rescue. Different noninvasive brain interfaces are now available that allow people to generate mental impulses, usually through physical movement of the limbs, that can be electronically transmitted (currently by phone) to someone with a similar setup and deciphered with a codebook that tells them what the impulses mean. It's slow. In 2016, the five-letter word *hello* was transmitted from the United States to London, and it took six hours. (I would imagine, however, that after the first three letters, even a nonpsychic could save a couple of hours and guess the stupid message.)

But it is progress. Maybe we should look at this as though it's in its dial-up phase and someday will get to the high-speed fiber-optic version, wherein you can put on your electrode cap and transmit your thoughts to someone else. Or you could just call them.

TELEKINESIS

Let me state for the record that there is no study, no evidence, no video, no nothing that says people can move shit around with their minds. And if there were, if it did exist, don't you think we'd know about it? I mean, this is superhero-and-supervillain stuff right here. That's why when it happens in fiction, it is either an origin story (easy to keep under wraps) or it takes place up Old Butcher's Holler or somewhere similarly remote.

No discussion on disproving psychic powers would be complete without mentioning the Amazing Randi. James Randi, born in 1928, started his career as a stage magician, but he soon realized that people not only wanted magic to actually exist, they also believed it actually existed. A regular contributor to *Skeptic* magazine, he established the James Randi Educational Foundation to investigate supposed claims of paranormal activity.

Perhaps the most well-known purported telekinetic is Uri Geller, an Israeli-born illusionist who went that extra mile and claimed that the tricks he was doing were not tricks, that he actually possessed "powers." During his heyday in the 1970s, his big trick was bending spoons by rubbing them with his fingers. (That was what passed for entertainment back in the 1970s, but believe me, this guy was huge.)

Randi was a regular guest on *The Tonight Show Starring Johnny Carson,* and although Carson was a huge fan of magicians (and an amateur one himself), Randi would come on not to perform magic, but to expose how charlatans pull the wool over your eyes. He showed how witch doctors use animal blood and chicken parts to pretend to pull tumors out of the sick. He showed how to use mirrors to peek at drawings you were supposed to be copying with your mind. And—when Uri Geller was on *The Tonight Show*—Randi consulted with Carson to put the psychic's powers to the test.

All the experiments performed for the show were things Geller had done many times, and the only difference was that his assistants were not allowed to set them up for him. So, not surprisingly, Geller failed. Spectacularly. Couldn't do a thing. Said he wasn't feeling it. But the real reason was that *it was all bullshit.* Hiyo!

If you google Randi, you'll even see him duplicate the spoon bending, which is so simple, it's infuriating that someone like Geller could build a career on it. You just bend a spoon ahead of time until it's weak enough that you can easily finish the job in front of your mark. Randi even made a long-standing offer of a one-million-dollar prize to anyone who could prove he or she had psychic powers. No one was ever able to claim it.

Besides, if you really had that kind of power, would you use it to bend fucking spoons?

GUH-GUH-GUH-GHOSTS!

In 2010, I took comedian Tom Green to the Whaley House in San Diego, supposedly a top-ten ghost joint in the United States. I booked the *Ghost Hunters* crew through my old friend Johnny Gumina, and they rigged the house to catch, in regular video and infrared, anything that moved. The plan was to spend the night, from dusk 'til dawn, looking for ghosts and scaring the (few) wits out of Tom.

Our ghost experts were a group of women who were clearly authorities on the subject, because they had written books about it. If you think *I'm* not a real professor, you gotta hang out with these ladies. The first proof they offered was to go into the next-door graveyard (convenient) and look for "orbs," little floating faces of the dead that show up on digital cameras. I took one picture and immediately had five big orbs, two with faces. But the experts told me that was just evening moisture in the air. You had to be "sensitive" to actually capture an orb.

Being "sensitive" turned out to mean "has had a neurological event." There seems to be a running theme wherein one can communicate with the dead after having had a personal tragedy involving the death of a loved one and a sharp blow to the head. It was all turning out to be pretty sad for a late-night comedy piece. But then . . .

Three a.m.: We were in the main room when the chandelier began to swing overhead. Our host turned off all the lights in the house to welcome the spirits. We crept upstairs, and—THE SHADOW OF A WOMAN IN VICTORIAN DRESS CROSSED THE WINDOW! We switched to the infrared camera, and there she was, plain as day, a woman in mourning dress pacing back and forth!

Tom was terrified. Me and Joe Medeiros, the head writer, were telling him what we saw on the monitor. We followed the woman into a small anteroom as she stepped on creaky floorboards and pushed past doors with perfectly nonoiled hinges, then slowly lifted her veil and . . . turned on the light. She introduced herself to Tom—big fan!—and explained that she worked for the Whaley House under the pretense that ghosts are much more likely to pay a visit if you dress up in creepy clothes and pretend to be one.

In the morning, we did an accounting. Our three experts had seen eleven ghosts and talked to two of them. We had seen a lady who works as a ghost part-time, when she's not managing an Arby's. And our crew, God bless them, decided at that time to pipe in: In all their years of rigging and shooting haunted houses across the world for *Ghost Hunters*, they had never once seen any evidence that ghosts exist.

TALKING TO THE DEAD

There are so many cases of séances and mediums and talking with the dead that have been proven to be hoaxes that it would take many books just to scratch the surface. This is perhaps the greatest case of wanting to be fooled, because the payoff—chatting with people you miss and being reassured that you're going to have an afterlife—far outweighs the fairly simple, explainable truth.

Here are just a few examples:

In the 1880s, the Fox sisters—Leah, Kate, and Maggie—claimed to communicate with the spirit world by eliciting a series of raps as responses to their questions. Turns out they were cracking their joints, but that didn't stop half the world from believing them. They became wealthy mediums until two of them confessed to their hoax, showing how they did it. They died five years later in abject poverty. FAIL!

Bill Wilson, one of the founders of Alcoholics Anonymous, was into the occult. He had a "spook room" in his house, and would frequently make contact with the dead, in particular a fifteenth-century monk named Boniface. Between Boniface and his other spook pals, Wilson credits AA's 12 Steps to help he received from the spirit world in that room. He even claimed to have proof positive that ghosts exist, but he declined to share that evidence with anybody, even his AA group, who were following the otherworldly program. So, this paragraph doesn't prove or disprove anything, but this anecdote always surprised me, and it gave me a chance to use *spirits* and *proof* without actually saying what I'm thinking.

Harry Houdini had his own James Randi–style prize for anyone who could prove they had communicated with his dead mother after she passed. A lot of people are confused about Houdini and the occult: Most think he offered this prize—and was so interested in séances and mediums—because he believed in all that stuff. But the truth is, early in his career, Houdini would hold fake séances and knew all the tricks. After his mother passed, he felt the exact kind of grief he had exploited in others, and he made it his mission to expose the frauds for what they were. It will come as no surprise to you that no one ever claimed the prize, which is still on offer: After his death, Houdini left behind an envelope that, if you could recite the contents, would entitle you to the money.

In most cases, "ghosts" can be exorcised not with a priest or a tiny lady chanting, "Welcome, children! All are welcome!" but with some hinge oil, flooring anchors, and pruning shears to stops the creaks, squeaks, and thumps that are ever-present in this mundane world.

SEEING THE FUTURE

There are more storefront psychics now than at any other time in history. They are scam artists. All of them.

In 2015, the *New York Times* ran an article about a program in the New York State prison system that offers incarcerated psychics a chance to come clean and get early release. I mean, there's a whole program! Just for psychics!

Their stories are all the same: Encourage people to tell you their problems, then cook up an expensive solution that will "solve" them. Usually, that means handing over some valuables or cash. Sometimes it means buying a talisman or doing something else that will profit the psychic. While this explanation is under dispute in the case of the Winchester Mystery House in San Jose, California, that seems to be the explanation for its existence: A psychic tells the heiress to the Winchester rifle fortune that she must keep building an enormous mansion to house all the spirits killed by the guns. And the psychic knows a great builder—her husband.

It's a lot easier to scam someone in person, and it helps to be observant and sensitive—not to the spirit world, but to clues the mark unconsciously gives you: Nodding of the head when you're right. Microexpressions that indicate whether you're on or off the right track. Eye contact. Body language. Breathing patterns. And that is why undercover cops and debunkers learn to control these responses and fool the scammer.

These psychics are preying on sadness and hope. And they can be ruthless. One potential parolee, Celia Mitchell, scammed a woman out of $160,000. And that was just one client. In every case, the psychics admitted to simply lying. Reading tarot cards is just a way of making general statements to elicit specific responses.

Perhaps the biggest psychic scam of all time was Miss Cleo's Psychic Friends Network. We are school friends with a family whose dad was the technician who set up the Psychic Friends Network hotline: He did nothing wrong, but a lot of those guys who ran the network went to jail after an undercover policewoman completed the classes that show the (simple) techniques that let you guess stuff about a person—and, more importantly, keep the sucker on the phone (at $1.99 a minute). Often, they simply tell a story that is true of everyone: "You try to be organized, but occasionally things fall through the cracks." "Yeah! You totally know me!"

At one point in the 1990s, Miss Cleo's operation, most of whose workforce consisted of teenage minimum wagers, was clearing more than a hundred million dollars a year. After facing false advertising charges, however, she was bankrupt two years later. She should have seen it coming!

Chapter 16

THE SCIENCE-FUL WORLD OF DISNEY

When you think of Disney, the first thing that comes into your head may not be science. You may think of sitting on teacup rides, eating Dole Whips at Disneyland, and the dozens of charming near misses that are supposed to be entertaining to kids, like that half-scale riverboat and Tom Sawyer's Island, or Tarzan's Treehouse, or the mind-bendingly slow submarine ride that is to theme park rides as 1980s electronic football is to *Call of Duty*.

You also might think about lawyers. *The Simpsons* wasn't kidding when it showed lawyers suing a church fair for calling itself "The Happiest Place on Earth"—these guys don't mess around with their intellectual property. Ever cautious, the waitstaff won't sing "Happy Birthday to You" at the restaurant in the Pirates of the Caribbean ride, even though a judge has cleared the song as public domain. When my son Dashiell was four, we were at Disneyland in the little New Orleans section, and employees were throwing beads from the balcony overhead (no breast exposure required). Overzealous teenagers started fighting the little kids over them, and a huge guy fell on Dash. First employee to the scene? A lawyer with a document and a pen. But we got free ice cream! Although it was Disneyland ice cream, which is kept on dry ice and is not soft enough to eat until the kids are home, asleep in bed.

To many, just bringing up the name Disney creates involuntary earworms. Just reading the words "it's a small world" puts me into a world of laughter, a world—hang on, I don't want to go too far with this. (See the "lawyers" paragraph above.)

Before we get into the actual astonishing Disney science details, I would like to touch on the pioneering social engineering that goes into many Disney products. Frederick Law Olmsted, the guy who codesigned Central Park in New York, designed and oversaw the landscape architecture for the Chicago World's Fair of 1893. He wasn't just creating a space for exhibits that showed everything from mock villages featuring indigenous peoples of various cultures to the astounding technological innovations of the day. He was designing an experience that allowed visitors to turn the corner and find themselves in a new land.

One of his workers, beguiled by the concept, would go home to his family every night and talk of the wonders of what they were building. That man was Walt Disney's father, and Walt would go on to use the concept to build Disneyland and Walt Disney World—not just theme parks, but user experiences that allow you to travel the world in a pretty small space. Think about that the next time you are at Knott's Berry Farm, where you can see all the way across the park as you're getting patted down at the entrance while watching thugs hide their weapons in the bushes. (Criminals love theme parks: Most weekend nights, the police recover several stolen cars at the Farm.)

Sure, it's easy to present Disney as an impersonal monolithic corporation that buys things like Marvel Comics and Lucasfilm whenever it feels like it. That may be true, but when it comes to building and working in the parks, the longtime Imagineers love it. They are devoted to it. I have met a few old-timers who delight in the wonder they struck into the general public, with big statements like the Hall of Presidents and little touches like the painted eyeballs that stare out at you from the knotholes in the woodwork. These guys affectionately call Disneyland "the Shop" and Disney World "the Factory"—and the respective nicknames are due not only to the size difference but also to the fact that Disneyland's Imagineers have more creative freedom.

I'll bust a myth about Disney World before we go on: It does not have a system of tunnels and underground spaces running under the entire park. Oh, it has a park-sized space under the park, but it's above ground. When Walt Disney purchased the Florida land, it was mostly a swamp, very expensive to drain and tunnel into. So, he simply built the first floor to act as the basement and worked his way up from there. This is just one example of how Disney solved engineering and design problems with ingenuity and elegance.

PEST CONTROL

Concerned that Disney is taking over your childhood stalwarts Marvel and Star Wars and ruining them? Or, to be more precise, making them better? Well, you'd better pivot that paranoid ray toward the actual ecosystem of central Florida instead because, in order to give millions of visitors a better experience, the Walt Disney Company is *playing with the very elements of nature itself!*

And by that, I mean they have cool ways of controlling the pest problem, because somewhere along the line, Walt decided to build his church not on a rock but on a swamp. Unlike the Trump administration, he both was able to drain that swamp and decided not to, as mentioned above. But building the basement level aboveground was only the beginning: Swamps have a lot of pests, some of which can carry diseases, and in the case of the bugs of central Florida, they are big enough to carry off young children. So, how do you push back nature without carpet-bombing the tourists with chemicals? Fight fire with fire! Or, as it turns out, bugs with bugs.

Meet the ladybug ranchers. No, this isn't what the bully ranchers call the sensitive ranchers; there are actual Disney workers who raise millions of ladybugs and release them into the park and the extensive environs of Disney World. These little beetles are not only miracles of flight engineering, with underwings that fold in a complex natural origami that is the subject of a new study at UC Berkeley, but, when you get right down to it, they are also gnarly little predators. They control the population of several other species of insect that would be annoying to guests or destructive to the plants and topiaries, like whiteflies, mites, thrips, and aphids, to name a few. And they eat other soft-body insects and their eggs, too. Will a

rampaging population of Disney-bred ladybugs take over the world? Let's hope so!

Of course, why stop at ladybugs? Disney World is also the happiest place in Florida for the awesomely named minute pirate bug—or minute pirate bug of the Caribbean, depending on where in the park you find one. Along with another creature called the big-eyed bug (which *should* have its own ride, then film franchise, then ride based on the film franchise, and, finally, a bug named after it), these two species eat caterpillars and other nuisance bugs. You can't spell *catering* without *caterpillars*!

Disney also employs a vast army of snakes—and no, I don't mean the aforementioned lawyers (although I really do). Because you need snakes to control the rat population (and I don't mean lawyers, although again, I do). Here's an experiment you can do on your next trip to Orlando: Once you're out of the park, take any tiny scrap of food, or even something shiny. Drop it on the ground. Within seconds, it will be covered by a seething blanket of nasty scavengers. By which I mean lawyers. But I specifically *don't* mean rats, because there is a distinct lack of rats on the Disney property, owing to friendly species of snakes like the black racer, which not only work pro bono but also eat the rats before they can get near your Dole Whip–dropping kids. And, come on, Disney, why is there no *Black Racer* film/ride/TV/ride/film franchise by now? Can't your lawyers get the rights?

PATENTS

If you feel like you could be doing more with your life, don't read on. Go get a drawing board, an electronic tablet, or whatever you do your thinking on, and start inventing. And even then, you will never get close to the number of patents the Walt Disney Company has been granted, because in the time it has taken me to type this, they probably just got five more.

Take a look at any of the patent sites, and you will see that Disney averages between fifteen and twenty patent grants *per month*, which seems impossible from a company that kept coming up with cornball portmanteau words like *Imagineers* and *innoventions*. But this is the real stuff, cutting-edge tech that not only enhances the user experience in the parks but also protects the Disney brand in a world of changing technology.

Of course, there is a long tradition of Disney innovation leading to patents and fortune. The very early animations that Walt did, before there even was a Disney company, featured the integration of a live-action actor in an animated world, a very intensive process that allowed him to create more cartoons faster than other companies and get a competitive advantage. He went on to invent a great number of innovations that improved the speed and quality of cartoons, allowing him to create the first feature-length animated features—most notably, the multiplane camera.

This invention used transparent animation cels you could see through in many layers to move backgrounds, foregrounds, characters, and other elements independently so he could keep up with live-action camera movements and techniques. (*Bambi* wouldn't be a cinematic masterpiece without it.) The whole thing stood more than two stories high, since a long-angle lens was required to compress the different moving planes and keep them all in focus.

I love animation—fans of the *Edumacation* podcast might be familiar with some of the things I've done with Kevin Smith, like building and animating the Bratzis in Kevin's 2015 feature *Yoga Hosers*. For that, I used Autodesk Maya, a professional-level 3-D package used at Disney and Pixar, though they also have their own proprietary packages. And Maya wouldn't be Maya if it wasn't constantly being improved, often through scripts and plug-ins written by Disney workers. Next time you see realistic eyeballs on a goldfish, you know whom to thank.

In just the last few months, Disney has been awarded patents for such cool and odd items as a free-form audio speaker that can assume many shapes, presumably for use in the parks to let them project sound from nearly anything. And a method of 3-D printing that would make the printed object unscannable and therefore immune to nefarious copyright infringement. My favorite is a system that analyzes human behavior and is able to predict when someone is about to take a picture, then optimizes animatronic characters to anticipate it and strike a pose in time for the photo. Very cool, but it also makes me sad that the robots in Disneyland will never be in a photo with their eyes closed, yet my mother always will.

DRONE DISNEY

If getting photo-bombed by a bunch of robots doesn't blow your hair back, take a look at the system of drone innovations Disney is planning on implementing in the very near future.

"Pixel drones" are essentially based on the same consumer drones that are making the beach even more annoying than it always was. But instead of three or four, literally thousands of these little suckers are controlled by a master system that makes them work together like a swarm of bees—if bees had lights attached to their bellies, able to emit colored light. Once it gets dark in the park, these guys all fly overhead like electronic bats, then use their lights to create 3-D displays that can show video images, form into 3-D characters that move and morph, and even simulate fireworks. (However, there is no plan to replace the Disneyland and Disney World fireworks show, I hear, even though Disney is the biggest purchaser of munitions in the United States after the military.)

Disney has also built a system of fewer—but larger—drones that carry a rear-projection screen into the air and can be positioned anywhere. One or more additional drones project onto the screen while even more broadcast the sound, creating a big-screen theater in the sky that can go anywhere. The projection can even include a scan of the night sky behind the screen, creating the illusion that whatever is being shown is floating in the sky. Forget the military drones in use today; just use this system to fly a menacing recording of President Trump over an al Qaeda encampment in Afghanistan, and they will all go running for the hills.

But that's not all. If you think marionettes—puppets controlled by strings—are creepy/funny (as in *Team America: World Police*) or creepy/creepy (as in every other kind of marionette ever), you're not going to like this one. Disney plans to send gigantic, five-story marionettes into the sky with a drone attached to each string, with a computer coordinating their movements to make the whole thing more lifelike. So, if the Wicked Queen doesn't make your kid spit dirt in his pants on your next trip to the Magic Kingdom, wait until a 50-foot-tall Jack Skellington shakes his moneymaker Gangnam-style in the night sky over Anaheim. Just thinking about it makes me want to curl up in a fetal position and wait for my Disney ice cream to warm up to an edible temperature. This whole idea is going to make the Electrical Parade look as primitive as, well, the Electrical Parade.

If the thought of bringing your child to a place where a battery of what are essentially four-bladed flying lawn mowers are swooping overhead, have no fear. The key invention that will be on all Disney drones is the Disney-patented drone airbag. If there is any kind of malfunction or the device gets too close to the ground, the Matterhorn, or the Dole Whip stand, it is instantaneously encapsulated in an airbag that will cause no harm as it plummets from the sky. Sure, it'll make Jack Skellington look like he had a stroke, but nobody gets hurt.

INNOVATION, OR JUST WEIRDNESS?

I can't write about Disney without including a speed round of systems, innovations, and details that are just plain weird. For example, in writing this chapter, I discovered that Microsoft Word automatically capitalizes anything with *Disney* in it. This is not true of the names of most towns in America, but, hey, it made my life easier.

Disney innovated crowd management—there are more things to see and do while waiting in line at Disneyland than there are attractions, restaurants, and rides combined. But when it comes to employees talking about crowds, they have a code. The best one is "code V," which means a guest has vomited, presumably because they just saw Jack Skellington having a stroke.

The most popular souvenir at the parks is the personalized hat with Mickey Mouse ears—but you can't have a celebrity name or a curse word embroidered on it. However, if you find a particularly green employee behind the counter, you can trick them: While we were writing *White Chicks* together, Shawn Wayans had *monkey* put on his ears, and he wore them the whole day at the park. I was offended.

Most of the water rides are on tracks. So, while that enthusiastic Cal State Fullerton student operating the Jungle Cruise ride is going through the same jokes guides have been doing since Ike was still carrying his own clubs, you're really just on a level track that's been flooded. Why don't you know that? Because they dye all the water green to hide the track.

All the plants in the Tomorrowland section of the park, by design, are edible. That's right—if the Dole Whip stand is too far, or you're still waiting for your ice cream to get to a state where your tongue won't stick to it, you can eat the plants like you're in the "Pure Imagination" scene of *Willy Wonka & the Chocolate Factory*. But don't eat too much, because you might just go code V. (I wonder whether George Lucas, who attended the opening of Disneyland at age eleven, took a bite out of Tomorrowland?)

While Disney World has ladybugs, Disneyland has feral cats. More than 200 wild cats live in the park, and the company has never schemed to kick the cats out, because they control the vermin, particularly rats. Park management like them so much, in fact, that in 2005 they gently trapped each of the cats, attended to their medical needs and neutered them, which may not sound nice, but it keeps them from spraying Mr. Toad's Wild Ride.

And finally, on a more science-y note, when they decorated the scenes in the original Pirates of the Caribbean ride, they used real skeletons obtained from the medical lab at UCLA to maintain a high degree of accuracy. While most have been replaced, there is still one real skull left—on the headboard in the pirate treasure tableau. If only my California driver's license had that as a donor option!

Chapter 17

SCIENCE FICTION
TO SCIENCE FACT

Here's my misspent youth: I spent many, many hours in a camping hammock behind my parents' house devouring the Science Fiction Hall of Fame series, everything by Arthur C. Clarke, many things by Robert Heinlein, the complete works of Philip K. Dick, and especially issues of the short-lived *Galileo* magazine. I also read *Asimov's Science Fiction* magazine, which was a lot like *Alfred Hitchcock's Suspense Magazine*, but with ray guns and the peculiar feature of having a different photo of Isaac Asimov in the upper left corner on each issue. One photo was of Asimov's feet.

I got into sci-fi when I was a young teenager. That's when I really started pouring on the steam as a competitive swimmer and a member of a barbershop quartet, which meant I had a lot of free time to read all the way through high school while other young men were spending all those tedious hours exchanging bodily fluids and viruses with young women (or each other). I was filling my head with possible futures, then thinking about them (or singing the baritone parts to one-hundred-year-old songs) for six hours a day in the pool. I never grew taller than 5-foot-6, which ruined my chances of becoming a world-class swimmer, and I never was able to grow a handlebar moustache, which ruined my chances of becoming a professional barbershop-quartet singer. (So, thank God.)

I was super into gear, and dreamed of possessing impossible inventions, like I was Miniver Cheevy longing for the Medici gold. A chronosynclastic infundibulum that you could activate under your arm, which would serve as a holster for whatever you wanted to carry and was completely undetectable to anyone in the universe? I wanted. A portal projector that would allow you to walk through any wall, could teleport you anywhere, and somehow regulated the difference in temperature and air pressure? On my wish list. A watch that you could set to go back and undo all of the embarrassing moments in your life? I still want that. In fact, if the Devil is reading this, write my publisher for my email and we can make a deal.

Of course, the coolest shit came in the dumbest stories. Bad science fiction is defined by constant stops and starts so they can explain the gear. "Tyrone Xandar pulled out his X251 Stratoblaster—a plasma energy pistol so powerful it could melt the metallic fur of a Denebian magmabeast"—stuff like that. Really takes you out of the story.

But there were so many things in fiction, movies, and TV shows that people just took for granted. Cool shit that was part of their world of the future. Keep in mind that I was a little kid in the 1970s, when everyone thought that polyester clothing and digital watches were pretty fucking awesome. I mean, if you didn't have to know what the hell the big hand was doing while simultaneously generating projectile BO from the pits of your tight plastic football jersey, it meant the future was something we controlled.

Most of the 1970s, from the Carter administration to *The Love Boat*, did not work out. But don't forget that there was a little thing called *Star Wars* that came along somewhere in there, and it carried on the tradition of using science fiction to tell stories that could have happened to anyone, in a future that had some pretty cool shit.

EARLY SCI-FI

Everybody looks back at Jules Verne for the predictions he made about the big stuff. For example, even though he didn't come up with the idea of submarines, he did understand how they would work and the danger they would present to shipping once they were able to stay underwater for long periods. (Side note: The title of *20,000 Leagues Under the Sea* refers to the distance Captain Nemo's *Nautilus* travels in the book—around 70,000 miles—not, as many people believe, the depth to which it submerges.) But Verne also foresaw the moon landing in *From the Earth to the Moon* (his 1865 novel, not the HBO series introduced by a stentorian Tom Hanks).

As you will see in this chapter, the trick for sci-fi writers like Verne isn't just to come up with the big stuff; if you're going to have a vision of the future, you gotta go for the cool ideas, too. And so, in a piece called *In the Year 2889*, Verne (and/or his son) predicted skywriting. He thought it would take mankind 980 extra years to get there, so he's a jackass, but let's not judge.

Aldous Huxley's 1931 *Brave New World* is not only the most fucked-up book you have to read in high school (or middle school, in my kids' cases), but in presenting a dystopian vision of the future that isn't strictly political, like that of *1984*, it's also just begging to get dinged for getting shit wrong. And Huxley did hang his hat on the Malthusian premise that population would outstrip food supply, which would be upsetting, hence (or perhaps an excuse for) all that weird sex stuff. One of the key elements to keeping the masses happy in the story was his invention of an opiate of a sort: soma. Actually, to be more specific, soma was an *antidepressant*. And even though (to our detriment) we have more food

than we know what to do with in our best of all possible countries, we take a *lot* of antidepressants. One in three Americans hold a prescription for antianxiety medication, and one in six have continuous antidepressant prescriptions. Which is depressing. What will I ever do about it?

H. G. Wells not only accurately predicted the 1938 invasion of Earth by Martians in Grovers Mill, New Jersey (which was later covered up—*shhhhh*) in his 1898 book *The War of the Worlds*, he also predicted the invention of the heat ray. "Wait a second," you interject. "Heat ray, you say? No such thing!" First of all, stop interrupting my book—it's hard enough hitting all these fake deadlines they're giving me, and second, yes, the US Army *has a heat ray*. It's called the Active Denial System, which sounds like another antidepressant but is really a microwave beam that makes you feel like you're sunbathing in Death Valley in July.

Victor Appleton's Tom Swift series will be remembered not only for its adoration of Thomas Edison and its alarming casual racism but also for inspiring the invention of the taser. Not only does the taser work almost exactly like Tom's electric rifle, it was *named* for it: *taser* is an acronym for "Thomas A. Swift Electric Rifle." So, the next time a protestor begs you not to tase him, bro, you can tell him that story while he lies twitching on the ground (I am assuming many of my readers will be riot police).

There are many examples like these from early sci-fi, but few predicted innovations that would have incredible, far-reaching impact on human behavior like Edward Bellamy did in his 1888 story *Looking Backward: 2000–1887*. There were no zeppelins, no cosmic disintegrators, no pultroonium-523 or molecular acids in his work. But Bellamy did predict credit cards. And for that I will never forgive him.

STAR WARS

THINGS GEORGE LUCAS GOT RIGHT

Landspeeders: Yes, if you're going to travel the Jundland Wastes lightly, why not do it on a cushion of air? Well, Aerofex of California has developed the Aero-X vehicle, which they describe as "a hovercraft that rides like a motorcycle." Although it's not really a hovercraft, which requires a smooth surface like water to keep the pressure going on its air cushion, and this thing can fly well above any surface. Aerofex has working models they drive around out in the desert, which, as I may have mentioned, is not to be traveled lightly.

But I will consider any developments to be prototypes until it can carry a driver, two droids, and a Jedi who controls stormtroopers' minds while making the asshole sign with his hand.

Lightsabers: You can't gut a tauntaun with one yet, but researchers at Harvard and MIT have found that if you shoot two photons through a cloud of supercooled atoms, they emerge as a discrete, single molecule that can be contained and manipulated. However, scientists at the Institute for Advanced Studies in Austin say that lightsabers will never be invented, because the process is impractical. What they don't say is that they are *Star Trek* fans, and that George Lucas ruined their childhoods with midichlorians and that awful *Crystal Skull* movie.

Droids: kind of a weird term, because it is obviously short for "androids," which technically means "automata that simulate humans in form and behavior." And also a trademark of Lucasfilm. So, C-3PO is an android with whom you would never want to go on a car trip, because he seems to be programmed to bitch about pretty much everything. But R2-D2? Robot, as are all R2 series astromech droids. IG-88? Android. IT-O Interrogator ball? Robot. So, the machines that weld our cars together now? Robots. And the androids we have as of this writing are not worth enumerating, because they are so creepy. Now, if you'll excuse me, my wife is gently asking me to furl up my nerd flag and get on with things.

Hyperspace: Before *Star Wars*, *hyperspace* was a term parents used to refer to playgrounds that have a vending machine with candy bars and caffeinated sodas. But then along came the *Millennium Falcon*, which, despite that cringe-worthy "12 parsecs" gaffe, was actually able to travel faster than light. Now, scientists and mathematicians at the University of British Columbia have developed a model that proves faster-than-light travel is possible. Sure, it's been theorized in the past, but these guys straight-up mathed us, son!

THINGS GEORGE LUCAS GOT WRONG

Midichlorians. Hayden Christensen. The shadows on Tattooine. Skipping voice lessons in middle school. The Gungan race, and one egregious Gungan in particular, whose name-name will not be mentioned-mentioned. What *parsec* means. The idea of a re-re-re-release. Ewok zippers. Brother/sister make-out sessions. Blue milk. Anakin and Padme cavorting in the fields. Giving a young woman a character name that spells out "pad me." *The Star Wars Holiday Special*. AAAANNDD . . . the Jar-Jar Tongue Sucker. Look it up!

STAR TREK

Food replicator. Universal translator. Tablet computers. Tractor beam. *Star Trek* was full of firsts when it debuted in the fall of 1966. There is so much wonderful optimism built into the premise of that show, from the Prime Directive (which was broken as many times as it was brought up) to the multicultural Team Benetton–in–space casting (ignore the accents, please) to even featuring the first interracial kiss on US television (though aliens used mind control on Kirk and Uhura to make it happen under protest). And Kirk got so much alien na-na that you know Bones had a special space-penicillin hypospray vial set aside for daily use.

And *Star Trek* was full of predictions, too, although I'm not sure how far they explored them as concepts. For example, they had a food replicator, from which crew and guests would request such exotic menu items as soup or Earl Grey tea. Give me a food replicator, it's gonna be surf-and-turf from the Palm with frozen gold-leaf truffle from the Savoy. But in the future, I guess people want to appear humble when talking to a food replicator. Nowadays, there are several versions of food replicators, the most recent being a pizza printer that will work as a vending machine and that NASA is reoutfitting for the International Space Station.

The universal translator was first introduced on *Star Trek*. It could take any language—even telepathy—and turn it into mid-Atlantic droning with heavy reverb. And you know that disembodied brains are thinking things like "Fifty *quatloos* on the newcomer," because even though they completely consist of brain tissue, they twitch around a little bit when they think. Nowadays, we have several translators, the most interesting of which are small handheld units developed by a number of Native American tribes to preserve their dying native tongues. However, Google Translate isn't there yet. I took the phrase "Is this my favorite puppy I see before me?" and translated it first into Pashto, then Basque, then Sindhi, then Amharic, then back into English, and I got "I was the doll I liked before I saw it before." It may not be as confusing and bizarre as the translation in *Arrival*, but Google definitely has some work to do.

Plenty of other cool ideas came out of the Star Trek canon, but we'll finish up with the holodeck, which was first seen on the 1970s *Star Trek* animated series but didn't really enter nerd consciousness until *Star Trek: The Next Generation*. I always thought the holodeck was both underutilized and not fully thought out, but they had other things to worry about, like constantly letting out Number One's uniform and the strange sadness Denise Crosby seemed to bring to her role. Why didn't they use the holodeck as a real-time, planet-to-planet communicator that would make you feel like you were talking to someone in person? No more space leave! What happens on the holodeck stays on the holodeck! Anyway, VR goggles are like the holodeck, according to VR goggle manufacturers, so let's move on.

THE JETSONS

The original *Jetsons* was on for only one season. *One season!* Twenty-four episodes that debuted on Sunday night, September 23, 1962. Most people watched it in black and white. And through the backward storytelling system of Hanna-Barbera's board artists, directors, and writers, it became what *Smithsonian* magazine called "the single most important piece of twentieth-century futurism."

They ain't kidding. Some people like to dismiss *The Jetsons* as simply a flip on *The Flintstones*, which had all the modern conveniences like garbage disposals and cameras, but inside each was a jaded animal that would manually perform the task and then say something like "For this, I went to college?" But *Jetsons* writers were looking forward, not back, and many of their futuristic lifestyle ideas were nothing less than great. Or at least prescient.

Like flat-screen televisions. All their screens simply existed on the show, and the characters took them for granted. And, yes, our children now shrug when you walk them through Costco as you marvel at the ridiculous 80-inch displays on display. But when we were watching reruns of *The Jetsons*, we were watching them on 25-inch screens—*the biggest TVs you could own*—positioned *on the floor* in giant wooden cabinets. The idea that a TV could even turn on instantly didn't show up until the 1980s. If you touched the screen, it felt like you were palming a bug zapper. Televisions had a weird electric smell and a warning on the back that said if you opened it, you could die. And the signal *traveled through the air as if by magic*. Now, the only thing our flatscreens don't do that the Jetsons' did is make a spacey noise as they retract into the ceiling.

A few *Jetsons* ideas have come to pass that were, frankly, stupid. Like robot house cleaners, which do exist—but, really, come on. Do we really want to spend more than a thousand dollars to get a machine to do something a nice person already does—and that takes five times longer? Or dog treadmills. They have dog treadmills now! This belongs in the pantheon of inventions for the lazy, alongside self-foaming liquid soaps. And the Apple Watch. *The Jetsons* essentially had Apple Watches, because in 1962 nobody knew what a terrible, douche-y, smug, brand-loyal waste of money they'd be. Unless *you* have one, in which case they are a good idea (said the man writing a book that will be printed on paper in 2018).

Oddest invention? Tanning beds. *Tanning beds!* Mr. Spacely takes a call while shirtless and smoking a cigar and getting an indoor tan. Of course, that scene was a send-up of how frivolous and silly many lifestyle inventions would become, since there is a sun outside and so forth, so take that for what it's worth.

JAMES BOND AND IAN FLEMING

We all know a guy who says that *From Russia with Love* is the best James Bond movie because it didn't have a ton of unbelievable gadgets, just a briefcase with a choke wire and some gold sovereigns. That, of course, is a stupid opinion to hold. For one, it is *one* of the best James Bond movies because Quint from *Jaws* has blonde hair and kills James Bond–lookalikes for fun like he's in *The Most Dangerous Game*. And for two, the gadgets from the Bond movies are super-awesome! Either one of these undisputable facts would label the guy I'm talking about as an asshole, but combined they make him a *super* asshole.

Author Ian Fleming was legit when it came to the spy business. He worked as an intelligence officer for British Naval Intelligence, most notably on Operation Goldeneye, which is where he got the book title and the name of his estate in Jamaica. He probably named the operation, since he was a bird lover and the Goldeneye is a bird. Want more evidence? The author's name of his favorite bird guidebooks was *James Fucking Bond* [fucking mine].

Fleming knew his stuff when it came to gadgets. Real stuff they had at his time in intelligence included tie-clip cameras (more astonishing is the fact that people wore tie clips), hollow teeth containing cyanide (actually, a Soviet invention), and the microdot, which anybody could make with an ordinary 35mm camera because it was just an incredibly tiny negative—a picture of a picture.

What imaginative bits of Bondiana have survived and flourished as real equipment? How about the exosuit, from the brewery scene in *Goldfinger*? Or the jet pack from *Thunderball*, which—while it was wired to a helicopter in the movie—was an actual prototype capable of 30 seconds of flight (working on *The Awesome Show*, I got to see a demo of the Zapata Flyboard Air that can stay aloft for 30 minutes). The fingerprint scanner in *Diamonds Are Forever* is now available on your iPhone, though it might be the most easily hacked security measure in history. And let's not forget the homing beacon also from *Goldfinger*: Do you have a Tile connected to your key ring yet?

When it comes to fantastic gadgets, however, no Fleming story does it better than *Chitty Chitty Bang Bang*. That damned car—in addition to tolerating those obnoxious children—was also a boat, a flying car that could hover, was remote controlled and self-driving, and even seemed to have its own artificial intelligence—all characteristics we see in today's most cutting-edge cars. However, for all that foresight, Fleming never put in any drink holders.

THE MISSION STATEMENT OF THE EDUMACATION PODCAST IS TO INSPIRE CURIOSITY IN OUR LISTENERS AND THE REST OF THE WORLD.

Obviously, based on the quality of questions we get for the Why section, this mission is far from being completed. To be fair, the Why section flies in the face of conventional advice and infotainment shows: The audience sends in their questions, and instead of asking an expert, I ask Kevin Smith. Who is baked.

But if I can achieve just one thing with this book, it would be to stop people from asking, "Why do birds suddenly appear every time you are near?" I get it. It's a Carpenters song from the seventies. Big hit. Rick Moranis sang it in *Parenthood.* And it is a question. But it says way too much about you and doesn't really inspire a meaningful conversation. *So quit it!*

I started my adult career working for the Zagat Survey, which rates dining establishments, and there I got to learn many methods of conducting surveys. It turns out that conducting a survey is a lot like conducting an experiment: In order to get meaningful results, you have to ask questions to which you have already predicted the answers. If you want to squeeze something meaningful out of the vox populi, you can't just ask something like "What aspect of a restaurant is most important to you?" The answer will always be "the quality of the food," because the person who will fill out your survey on restaurants will always consider himself to be a foodie.

And foodies are a pain in the ass, because (except for a few extremely rare cases) they consider themselves experts on everything having to do with food. Do they have extensive Cordon Bleu training? Did they intern with Gastón Acurio? Or do they simply spend a significant portion of their disposable income getting someone else to cook for them?

So, you have to ask the same question in different ways. Try "What's more annoying? Mediocre food, service, or decor, or high prices?" They will always answer "mediocre service"—because, as we learned at Zagat, when you're at a restaurant, the most important thing is that you are served well. Not the quality of the food!

We also learned that letting people ask their own questions was one way of getting meaningful commentary. It says a lot about trends when you get 500 copies of the same question in the same year's survey. In the late 1980s, that question was, "Why don't the waiters tell you the price of the specials?" We told this to our restaurateur friends, and they started having their staff include the prices in the descriptions of the specials. Everybody wins!

By the time I left Zagat, the New York survey was getting around 10,000 responses, which is a huge sample. The questions I get on Twitter about *Edumacation* amount to about 250 a week—not quite the same, but about as many as you'll find in a Quinnipiac poll on the nightly news. Of that 250, I regularly get the same questions, and I am asserting that it says something about society when you look at the most common ones (I am paraphrasing): 1) "When will computers/robots be smarter than us and take over the world?" 2) "When we cure all diseases, will the only way we die be by accident and violence?" 3) "When do scientists think we will be hit by an asteroid and wiped out like the dinosaurs?" 4) "How much weed does Kevin smoke?" And 5) "Why do birds suddenly appear every time you are near?"

Let's toss out the last two and focus on the first three. There's a theme there, and that theme is all about death. And if you remember that buzzards are birds, the same goes for question 5.

The following section deals with some popular questions, some questions that turned out to have answers popular with the listeners, and some questions I wish the listeners had asked. And this time, I get to answer them!

Chapter 18

GAME CHANGERS

I have always loved games, in almost any form. In the early 1990s, newly married to Johanna, she and I were lucky enough to have a friend who worked at *Games* magazine, back when Will Shortz was editor. Our friend Lou was a game reviewer, and he would bring over new board games and such for us to play so he could get our reaction. We got to play some winners—Taboo notable among them—as well as some unbelievably complicated and forgettable games.

I also used to shoot a bit for Jay Leno called "Pitch to America," where inventors, TV producers, and yes, game designers would pitch their idea straight to the camera, and it was up to the audience to guess whether they had managed to sell it to a distributor or not. The point of that bit was to show that even if you think it's a good idea, there are a lot of things that can stand in the way of getting your idea out into the world. And some ideas are just stupid enough to get made and be successful.

A favorite fact that comes out way too often is that many games that involve kicking a ball are derived from ancient battle victories, wherein a loser's head would get kicked around by the winners. While I am sure that this did occur in some form, it's not like there were professional teams that always played with heads instead of balls. Severed heads are hard to kick, you get blood and brains all over your shoes, and it's not like you're going to be able to kick a field goal with a head—you'd break your toe (although kickball would be much more manageable). And it would explain why, in soccer, you can't touch the ball with your hands: Who would want to?

What's more likely is that people were playing ball games all along, usually with an inflated animal bladder or intestine, which was stiffened and protected with a cured hide as a cover—which is pretty close to what we have today. Footballs, many soccer balls, and some basketballs, to name a few, have a rubber air bladder surrounded by animal skin. Still, you don't get a lot of PETA folks chucking blood at LeBron James, although for a while there you might have seen Cleveland residents trying it.

Here, we look at games that we take for granted and for the most part know the rules to, but that have a strange origin, used to be something else, stand for something creepy, or are just plain stupid.

CHANTING GAMES

Some of the most horrible game origins come in the form of children's dances set to chants and nursery rhymes. Now I can't watch kids play any of these without thinking they're practicing to audition for a remake of *The Craft*.

"London Bridge Is Falling Down" is usually performed as a circle dance so that two kids face each other and form a bridge with their arms, and the other children walk underneath until the end of the song, when the bridge comes down and captures a kid. The fun way to look at this is that from the time of the Roman occupation of England and forward, London Bridge has been in disrepair at times, inspiring a drinking song to that effect. But the more sinister version refers to a tradition, for luck, of burying a child's body in the pier of a newly built bridge, and to the fact that when it was almost falling down, another child would be needed to keep it standing.

"The Farmer in the Dell" is a game that lets kids practice cruelty, so you might have guessed it comes from Germany. In it, kids stand around in a circle, and one is picked to be the Farmer. Then he picks a wife, and they pick the children, then dogs, then servants, then cheese. Which teaches you not only the feeling of being picked last for a team but also that servants are in a category just below dogs, and if nobody likes you because you wear glasses and might be a bit too enthusiastic about answering questions in class, you may be asked to play the role of the cheese.

"Ring Around the Rosie" takes the cake in my book. The lyrics could instead be "La la la la, dead!" as many folklorists believe the song evokes the horrible wait, during the 1665 Great Plague in England, between when someone in your house dies of the bubonic plague and the guy with the meat wagon comes to collect the body. As the body began to decompose, members of the household would line the clothing and stuff the pockets full of flowers, believing that the foul odor spread the disease. The "ashes" may refer to making lye by leaching ashes from the fireplace and using it to clean the bedding and sick area, or it might be from the practice of burning down houses that contained multiple victims (which would just evict the rats, who carried the fleas that carried the disease). And, of course, "all fall down" means we all die! Die! DIE!

Or, it could mean something else. Kid songs are fickle like that.

PLAYGROUND GAMES

Setting aside the mind-bending terror of being the smallest kid in school when they were picking teams in gym class, there are many bizarre and macabre elements to the classic playground games we all know. But they're for kids, so they're corny, which makes them corn on the macabre. For that reason, I will describe these games for those of you raised on *Angry Birds*.

Jacks, the game with ten metal spinners and a rubber ball, has many ways to play, but the basic idea is that you toss the pieces on the ground, then you bounce the ball and pick up the pieces between bounces, usually with the same hand. However, they were not always metal pieces; they were originally bones, usually the hock from the ankles of sheep or pigs, or occasionally the spine bones. (The game is also called knucklebones, although you don't use those particular bones.) Maybe the game was invented by kids with nothing to play with but the bones of dead animals bleaching in the sun, but it has been theorized that, as much play among children is imitative of things they see their parents do, it was innocent mimicry of the way fortune-tellers and shamans use bones to see into the future.

Tetherball, in addition to being severely disadvantageous to smaller folks who wear glasses, are too enthusiastic in class, and are still smarting from recently having been chosen as the cheese, is another one of those games that could have a nice origin—but probably doesn't. One theory is that it is meant to celebrate maypole action, the practice of wrapping ribbons around a big phallic symbol as a courtship ritual and to celebrate fertility in the spring. (Oddly, a certain *Dogma* director's daughter goes to a school that has a father/daughter maypole dance, so *bleeech*.) Based on human history, however, another theory makes more sense to me: You would tie body parts of vanquished foes to that pole and start batting them around. Much easier than trying to place-kick them.

Blind man's bluff is pretty creepy when you realize that it started out as an adult party game. Basically, you blindfold someone, then they walk around the party "accidentally" feeling up whoever and whatever they could get their greasy mitts on. For this reason, it was a favorite game of King Henry VIII, who broke with the Catholic Church just so he could clap those meaty paws on more than just his (first) wife. But even if that's not the motivation, the modern game does teach children to single out and taunt someone with a disability, even if it is a temporary one.

BOARD GAMES

How we loved to spend countless rainy days playing the closetful of board games we had in the old house in Philadelphia! But it doesn't rain in Los Angeles, so my kids just watch TV.

The Game of Life taught us many things: In life, you will drive a convertible, your family members will be shaped like dildos, everything (actually, not that much, when it comes down to it) happens by chance, and in the end, you will finish either in a mansion or the poorhouse. While this mostly describes the First Family, it has little to do with how life really is. But the game, in its original 1860 first edition, got a little bit closer. In that version, you could end up in disgrace, ruined and in poverty, or pursue a life of crime and go to prison—or, best yet, you might have to commit suicide! Not for reals—just in the game. I guess in a time when you could ship out with the navy at the age of seven, you had to be a little "edgy" to make kids want to play your stupid board game.

Clue, the deductive-reasoning game that introduced children to a world where death could come at any time, in any form, in any room of your house, was first created by an air raid warden in World War II London, where death could come at any time, in any form, and completely obliterate all the rooms in your house. His version was a bit on the nose: He named it Murder! and despite the exclamation point, it was never made into a musical. And before it became the more mild-mannered Clue (or, inexplicably, Cluedo in England), it had a much more fun arsenal of weapons, including a syringe, a vial (or, inexplicably, a *phial* in England) of poison, an Irish club, and a bomb. Yes! A bomb! "Here you go, kids, have fun, and self-radicalize!"

Monopoly has to be the most self-serving, morally dubious board game in history. Yes, I said *history*, because (after chess, which made its debut when the world's population was much smaller overall) it is the best-selling board game of all time. Who doesn't want to be a fat-cat moneybags who sends his playmates to the poorhouse, to jail, or worse by overbuilding and jacking up the rents? Well, true to the game play, the origins of Monopoly are just as much of a rip-off as trying to stay overnight at Park Place with two hotels on it. Even though the official Milton-Bradley version of the story completely ignores its origins, the game was first invented in 1902 by Elizabeth Magie. She called it the Landlord's Game, and it was meant to show *the absurdity of economic privilege and land-value taxation.* What's more, it had a second round meant to warn players of the evils of capitalism! It's irony in a box!

THE OLYMPICS

I get it. The Olympics doesn't just celebrate who's the best at exercising. It is also a metaphor for the skills that helped build kingdoms. That's why the marathon event (now somewhat standardized at 26.2 miles) originally was 25 miles and change: It celebrates the historic run of the soldier Pheidippides from a battlefield near Marathon, Greece, to Athens in 490 BCE (although back then they just called it BC). He ran all the way home!

But there have been some stupid events in the Olympics. And today I celebrate them instead of the ones that make sense.

If I could travel back in time, it would be to go to the 1904 Olympics in St. Louis, Missouri. Or "Misery," as many of the athletes must have called it, since it was miserably hot and nearly half the marathoners didn't complete the course, because they almost died of heat stroke and dust inhalation. Better yet, the supposed winner of that race was disqualified for riding most of the way in an automobile. And the real winner was an English guy kept (barely) alive with sponge baths and a concoction of brandy, raw egg, and strychnine. Holy shit!

But, oh, the stupid events. The 1900 Summer Games in Paris included obstacle-course swimming, which seems like old-fashioned skate-and-barrel jumping after the ice melted. Or the plunge for distance at the 1904 Summer Games in St. Louis, wherein you dive in the water, and the one who glides furthest without swimming is the winner. (What a nail-biter that must have been! And the training? Forget about it.) And what about the rope climb? You climbed a forty-nine-foot rope and were judged for speed *and* style. Style? What, you had to put some shimmy in your shinny?

There were other Olympics with events that immortalized their years. Skeet shooting is still an Olympic event, but until 1900, they shot at real pigeons. Hey, I lived in New York for fourteen years, so I'm no fan of those rats with wings, but I still don't want to see their guts blown all over the place between Timex commercials. Trampoline became an event in the 2000 Games, briefly raising it from its normal status of "redneck pool." They replaced the projectile shooting in the pentathlon with lasers beginning at the 2012 Summer Games in London, so evidently, we'll be seeing more planetarium docents in the games. BMX biking has been an Olympic sport since the 2008 Summer Games in Beijing, and they haven't had that many radical dudes at the games since Munich.

They even used to do a tug-of-war, which was unpopular, as it did not have its traditional accompaniments of the egg-and-spoon and potato-sack races.

Clearly, the greatest stupid event of all time actually made it through three Summer Olympics, from 1984 to 1992. It is the only oxymoronic sport besides "Pittsburgh Baseball." So grab your nose plug, press Play on "Nadia's Theme," and get ready for solo synchronized swimming. Was the International Olympic Committee worried that regular synchronized swimming was too hard to follow? (Synchronized diving, while also somewhat stupid, has two divers synchronizing to each other.) If you're solo, what the hell are you synchronizing to? I mean, who's to know if you screw it up? As a former lifeguard, I would dive in on most of the routines, because they look more like someone's having a seizure than participating in a freaking *Olympic event*. It's like one of those old-timey Hollywood swimming extravaganzas—but with only one woman diving into the pool. I feel like offering a prize for anyone who can prove to me that this is an actual sport. It can't be done!

Chapter 19

WEIRD WORD ORIGINS

I love weird etymologies. I love that common usage turns often-horrible words into everyday expressions that would have time travelers from the past thinking we are more barbaric than we actually are. I love that weird word origins apply to common words that everyone knows, so it's not a case of vocabulary-related elitism. I love that they are the most requested segment for the *Edumacation* podcast, because—let's be honest—they are the ultimate in cocktail party–factoid fun.

They can be Ellis Island words, like *checkmate* in chess, which originated in India—and when you win, you are not asking to pay the tab in Australia. You are saying, *shak mat*, which means "I win." If it was within the Indian tradition to add "in your face," that would also be part of our chess-related lexicon. As it is, somewhere along the line, someone was disinterested or arrogant enough to hear some dude winning a match and just went ahead and thought, "Poor devil's saying 'checkmate' for some reason, what what!"

They can also come from interlanguage mash-ups. My father-in-law was born and raised in San Antonio, Texas, and in addition to being law partners with former congressman Maury Maverick, earning a Silver Star in Burma, and defending conscientious objectors during the Vietnam War, he was quite the local historian. He explained that many words we associate with the western frontier come from the cooperation between the German settlers in south Texas (I'm looking at you, Fredericksburg) and the Mexican cowboys working the cattle in the vast ranches south of there. So, for example, *buckaroo* comes from a German dude trying to wrap his Teutonic mouth around the Spanish word for "cowboy": *vaquero*. (And until 1982, *buckaroo* had never been followed by the Japanese word *banzai*.)

So, some words are born great, some achieve greatness, and, in the case of nautical terms that infiltrate our language, some words are thrust upon us. From time immemorial, sailors before the mast (meaning nonofficers) were often "pressed" into service. That is, gangs of deputized goons would hit the shore and start roving around with clubs, grabbing any able-bodied men and kidnapping them to serve aboard His or Her Majesty's fleet. Living in the tiny wooden world of the tall ships had a language all its own, and since, at times, more British subjects were at sea than on land, seafaring terms and colloquialisms have had a profound influence on the English language.

However, now that the *Oxford English Dictionary* has added words like *selfie*, has added to the definition of *literal* the sense of "figurative," and is planning on including emojis in future editions, the whole shebang is about to be tossed into a cocked hat.

SAILING TERMS

"By and large"—in addition to describing Louie Anderson—is an expression that has come to mean "everything considered," but comes from the way you set sail in a tall ship. You sail "by" the wind by laying a course at an angle to the direction the wind is blowing (mostly into the wind), thus using triangular sails. The square sails come into play when you sail "large," or with the wind to your back. Taking both into account combines *by and large*.

"Cut and run" may sound like the silent crop dusting of a lunch table on your way to drop off your tray, but in sailing it refers to severing the cable (which has come to mean a metal cord, but in the old days was just a heavy rope) to the anchor so you immediately set sail and get out of Dodge without delay. You lose your anchor at the same time, which is a (ahem) heavy price to pay, an idea also conveyed in the modern phrase.

One of my favorites is "the devil to pay," because it's so misleading and it has the DEVIL in it. Although it doesn't. Because for many years, wooden ships were made watertight by painting, or "paying," hot tar into the caulked seams between boards (it's why sailors were called "tars," as in *Raiders of the Lost Ark* when Sala quotes Gilbert and Sullivan by singing, "A British tar is a soaring soul . . ."). The paying was done on the underside of most of the surfaces, which would keep the pitch from melting in the sun and from being tracked all over the nice, clean deck. That said, there was one seam that was particularly hard to get your boiling-hot pitch bucket, brush, and body close enough to paint, located where the hull meets the deck. So they called it "the devil's seam," and used "the devil to pay" when referring to a dreadful, dirty job that had to be done.

"Loose cannon" refers not to Nick Cannon in his bachelor days but instead to the dangerous event when one of the long guns broke free of its ties and preventers and started rolling around the deck. A heavy barrel could easily crush a man against a bulkhead or mast while moving only a few miles per hour, so the phrase has come to describe a person who is both dangerous and unpredictable—like Nick Cannon was in his bachelor days.

In modern times, "slush fund" refers to money earmarked for something unnecessary so as to be actually used for unforeseen (or secret) needs or activities. Like during Watergate, when the Committee to Re-Elect the President, which had set some money aside, ended up using it to pay the men who did the break-in. It comes from the way food was stored and prepared on long voyages.

Most meats were heavily salted to preserve them, and to get the salt out, they were first soaked in fresh water, then boiled before serving. Not only did this produce an awful meal that actually sharpened your teeth like stropping a razor, it removed most of the fat, which was skimmed as "slush," stored, and sold by the cooks when in port.

FOREIGN WORDS

Avocados are nature's mayonnaise. Each one is an omega 3–packed gift from nature just waiting to be smashed up and spread on toast. But you may want to think twice before crushing your next avocado: While the word itself means "advocate," it is really a bastardization of the name for the fruit in the language of the Nahuatl tribe of southern Mexico: *ahuacatl*. Which means "testicle." So, at some point in history, someone looked up at an avocado growing on a tree and said, "Hey, that looks like a bumpy nutsack." And no, *guacamole* does not mean "bashed-up nard paste." Though it should.

Namby-pamby, which has come to describe the condition of extreme wussiness, actually comes from an all-time literary dis that stuck. During the first half of the eighteenth century, an effete, waistcoat-wearing, sentimental snuff-sniffing writer named Ambrose Philips would write what was essentially the worst combination of daily affirmations and Hallmark greeting cards. And yet he was popular. But along came he-man poets Henry Carey and Alexander Pope (who gave us such balls-to-the-wall quotations as "To err is human, to purr is feline") to take that fucker Ambrose Philips down a peg. And, oh, did they. They nicknamed him "Namby-Pamby," which rhymes (because they are poets, after all) and has come to describe anything fey, like the way Mr. Burns throws a baseball.

Quack as a description of doctors and medicine might be losing popularity as a slang term, but I intend on bringing it back, much as I brought back *douchebag* in the early 2000s (you're welcome). Anyway, the term comes from an archaic Dutch term for a snake-oil salesman. "Hawker of salve" was expressed as *kwakzalver*, which was shortened to *quack*. Now, with WebMD and Dr. Tattoff, who was literally born to remove tattoos, the idea of a doctor being untruthful and acting in his or her own self-interest is most people's impression of all doctors, so why narrow it down by calling only some of them quacks?

Sabotage has a few meanings in our society. It can denote the first song that enabled many white people to get into hip-hop and not have to just pretend to like it. But let's stick with the traditional meaning, which describes the act of jimmying up the works somehow. Like, "The North Korean missile fell into the ocean due to sabotage by foreign agents." Or "I would have gotten an A on my presentation if Tom McCleary hadn't sabotaged it by making fart noises in the back of the class." This is another word of Dutch origin: Those wooden shoes that are always clogging up (see what I did there?) the sidewalks in the Netherlands? They're called *sabots*. And saboteurs would throw their shoes into factory machinery to stop it from working. And then they must have been caught, because they were the ones running from the scene *with no shoes on.*

ANIMAL NAMES

Ostriches are some of the biggest assholes of the animal kingdom. I gave a triumphant cheer during *Dude, Where's My Car?* because, for the first time on film, they were not depicted as fluffy, fun-to-ride, quirky mistakes of nature with enormous eggs. No, that movie got it right, showing ostriches to be the utter dicks they are. So, when you learn the origin of their name, it comes as no surprise it's short for the Greek *strouthokamelos,* or "camel-sparrow," because sparrows are dicks and camels are bigger dicks. Sparrows shit on your car, camels spit in your face, and ostriches whack you in the head with their own otherwise useless heads. Maybe they're mad that they can't fly; I don't care. Fuck ostriches, fuck sparrows, and fuck camels, for that matter.

Lemurs have it rough. They live in Madagascar, which has some of the most dangerous predators and environments in the world, and which means they will be forever embarrassed by their depiction in the movie *Madagascar.* Good lord, when you come out looking dumber than the penguins, that is some rough treatment. But when you look at the origin of their name, you realize they are badasses from way back. They were named by none other than Carl Linnaeus, the Swedish botanist who came up with the whole system of naming pretty much everything in the world. This guy had seen it all, but when he first clapped eyes on these little primates, he must have crapped his pantaloons, because *lemur* is Latin for "spirits of the dead." Either that, or the dead people in his family tree must've been quite spry and hairy.

I love the origin of *orangutan* for two reasons. One, my eighth-grade World Cultures teacher claimed they were called that because their fur was orange and their skin was tan, which is not what their names mean in even the tiniest way. Two, the name actually comes from the Malay and Indonesian words *orang*, meaning "person," and *hutan*, meaning "forest." So, *orangutan* means "person of the forest." Under that definition, was the Unabomber an orangutan? Do other Malaysians who live in the forest self-identify as orangutans? Was *Forrest Gump* released as *Hutan Gump* in Indonesia? So many questions.

We all know that *Dumbo* means "large-eared flying elephant with an incredibly well-protected copyright." And we all know that Dumbo's dad was Jumbo, and that name means "big." Ahh, but does it?

Many word origins are highly disputed, and my favorite among these is that of *jumbo*. It applies to everything from jets to American-sized portions. But that meaning is not the word's origin. Many sources claim that it came into use because that was the name of the enormous African elephant that toured with P. T. Barnum's circus until its death (and then for three more years after having been stuffed). But other sources claim *jumbo* is from an obscure West African dialect and just means "elephant."

If you want to get weird with it, refer to the *Oxford English Dictionary*, which traces the word's origins to *mumbo-jumbo*, which either was coined as a racist reference to incomprehensible languages (to the British explorers, who also said *mumble-jumble*) or may refer to a grotesque African idol called Maamajomboo—which I am hoping is the source of the modern derivation "Bad Mammajamma."

GRAB BAG

I want to toss a few more word origins in here. After all, many books have been written about word definitions and origins. They are called *dictionaries*.

Disaster underlines our pagan belief that the stars can influence events in our lives, which started in the ancient past and continued up through Miss Cleo and the Psychic Friends Network. As it still does: *Dis* means "bad," and *aster* means "star," so *disaster* means "bad star," and *inculpamaster* means "the fault in our stars."

Do you ever get super-angry when some British guy sneers at our use of *soccer* because they call the sport football? Well, you need to sit down and breathe through your nose, because it's really not that important. What *is* important is that we get *soccer* from the sneering British themselves! It's an outdated nickname they came up with for "association football"—I guess from that middle part of *association*, which doesn't make sense. But they're the same people that call St. George "Sinjin" and Marmite "food."

The word *nice* is hilarious, because we get it from Middle English—only back then, it didn't mean "well disposed toward other people." It meant "stupid." And it meant "stupid" from way back, all the way to ancient Latin: *Nescire* means "not know," and *nescius* means "ignorant." This tiny bit of knowledge has many uses in today's world. Like, if you were to say, "Now, that Donald Trump, he seems *nice*," you would be referring to a word definition that stretches back millennia.

Lukewarm is a nice word in the sense of the above word origin, in that it is stupid. The Middle English word *luke* means "warm," so *lukewarm* means "warm warm." It's like "the La Brea Tar Pits." In Spanish, *la brea* means "the tar," so the whole thing means "the the tar tar pits." Which sounds nice, by which I mean stupid.

Tragedy means "goat song." It's from the Greek *tragos*, for "goat," and *oide*, for "song." (That's where we get *ode*.) And if you've ever seen a YouTube clip of a goat singing, it's pretty freakin' tragic. If a goat song had words, they would be, "I hate being a goat." But they are funny, thus proving the axiom "Goat song plus time equals comedy."

Hazard comes from the Arabic *al zahr*, which means "the dice." I guess getting close to gambling when you're on a crusade, killing people in the name of the Bible or whatever, is somewhat of a hazard, as is playing said dice and losing your hair shirt in the process. In another example of the cosmic wheel turning toward coincidence, this may explain why the hazard-light button in your car is located *just below your fuzzy dice*! Believe it or not!

Loophole, like you might find in a tricky contract written by tricky lawyers to trick you, does not come from the law at all. Instead it refers to a hole in the wall of a castle or battlement, through which archers might shoot you in the face with their arrows. Just like lawyers use loopholes in contracts to shoot you in the face with their metaphorical arrows. The word *lawyer* itself is another example of a word origin not unlike *loophole*: in modern parlance, *lawyer* has given us the word *asshole*.

Nightmare is easily the most metal word in this chapter, but then again, this chapter has *namby-pamby* and *lukewarm* in it, so that's not saying much. The origin of the first part of the word is easy to trace, since *night* means "night." The second part? It's from the Old English word *maere*, which means "incubus," and the whole thing was put together in Middle English to denote a female evil spirit *that lies on you and suffocates you in your sleep.* So, it's not only metal, it's also meta, because knowing the origin of *nightmare* can give you nightmares.

Chapter 20

PARTY ANIMALS

There are a lot of theories as to why humans like to get fucked up. It is not enough for scientists and therapists to say that it feels good, because the side effects of drug and alcohol use can be painful, addictive, and devastating. However, it is enough for most regular folks who don't mind tearing it up at a wicked rager now and again.

One big theory is religion. It is clear that drugs play an important role in the way folks worship: From communion wine to peyote buttons, altering one's mind is often associated with having some transformative mind-altering—or religious—experience. Also, "opiate for the masses" is an appropriate term, as conforming to common religious worship makes people cooperate better.

Another theory is that drugs are just a shortcut to the end of the reward path. Think about natural highs: endorphin release during and after rigorous exercise, or the high of an orgasm after an expensive dinner and a ride around the park. When you get high, you cut out all those hurdles and get right to the buzz.

This theory is often supported through animal behavior, as many physiological explanations for human behavior often are. This way of thinking is a double-edged sword: It assumes that animals don't have souls and emotions and all the supposed higher-level brain function humans do, so if they like to get high, it means we do it to respond to our baser nature. And it is reductive in terms of how many folks have to deal with the emotional and physical fallout of drug use.

That said, it is pretty hilarious watching a cat high on catnip. And even though it's outrageous and wrong, the early days of television got a lot of mileage out of letting monkeys smoke cigarettes. And if animals weren't into partying, we never would have had the Great Wayne Squirrel Party of 1979.

To explain: My dad was an organic gardener ever since he was a kid. And when we moved to Wayne, Pennsylvania, when I was in middle school, he tilled himself a very large garden in the backyard. However, the community of squirrels in our new neighborhood was particularly destructive, and they were too smart to step into the Havahart traps (for rendition to the golf courses of Bala Cynwyd). So my dad, in his usual manner, devised a cunning plan, and I was his helper. But it was up to me to figure out what the hell was going on.

First, we went to 84 Lumber and bought an apple-cider press. With that in the trunk, we went to Lancaster to a U Pick Apples orchard and picked 10 bushels of apples. The next stop was the medical supply store, where we bought some 5-gallon jugs. We got home at two in the afternoon and assembled the cider press, then we pressed 14 gallons of cider (we overbought when it came to the jugs) and, batch by batch, discarded the pressed apple waste in the compost heap no good organic gardener goes without. At approximately seven o'clock, we finished a long day of work. I still didn't get it.

Then, in the suburban twilight, Dad ripped open a package of yeast and threw it on the compost pile. We forgot all about the apple juice.

For the next few weeks, our block was dominated by fat, drunk squirrels. They would stagger across the lawn right into an open cage. I pulled one off the side of a tree at about head height (I wore welding gloves); he was asleep, with his claws embedded in the bark. Another one fell asleep in our squirrel-proof feeder, proving that even a drunk squirrel can outwit the best birdhouse engineers we have to offer. And even though their pals were disappearing right and left, others still came back for more.

So, yes, animals like to get fucked up. Let's have a look, shall we?

DOLPHINS

Dolphins have incredible press agents, because everything you hear about them is that they're super-smart, and love each other, and live in families, and team up to fight sharks, and can diagnose cancer in people using their sonar like ultrasound. But they're not just the rock stars of the deep: They like to *party* like the rock stars of the deep.

Not only do they like to masturbate themselves on people and oars and pretty much anything else that's available, dolphins have also been documented in several videos carefully nibbling on puffer fish, self-administering the tetrodotoxin the puffer fish uses for defense. Which is crazy. What's more, they often do it hanging around in a little dolphin circle, and *after they've taken a hit, they pass the puffer fish around* like they're at some kind of subaquatic Grateful Dead concert. Or Phish—I should've said Phish.

Tetrodotoxin, 1,200 times more poisonous than cyanide, is deadly to humans. (And chill on the whole poison vs. venom thing, because it's both.) There's

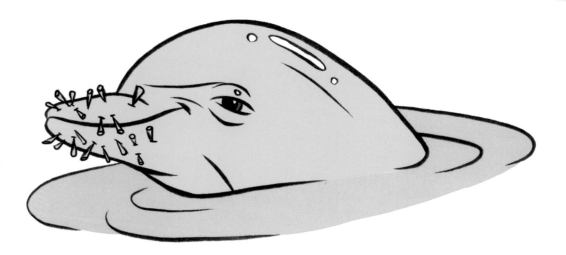

enough toxin in one puffer fish to kill thirty adult humans, and if you want an antidote, you're going to have to ask the folks in the next life, because there isn't one on this planet.

Do they do it to gradually accustom themselves to puffer fish toxin so that if they ever do get poisoned for real, they can survive? No. Do they do it on a dare, to build connections with a peer group? No. Do they do it to check out from the humdrum daily life of living in the ocean and switch their brains over to side two of *Dark Side of the Moon*? Yes!

CAPUCHIN MONKEYS

If you were a Capuchin monk, part of a Franciscan sect living in sixteenth-century Italy, you might have wondered what it was all for: praying, copying text, making beer, and atoning day and night. But if you could look back at it all, you'd be psyched to learn that not only are the four-dollar cappuccinos at Starbucks named after you, they also named the South American capuchin monkey after you because its markings resemble your monk robes. And that particular monkey likes to get *fuuuccked uuupp*.

To be fair, the monkeys get high in part for protection. They use several species of millipedes—whose own natural protection is to spray venom—and rub that stuff all over themselves. It wards off parasitic insects, but it also works like a narcotic for the monkeys. Most of the varieties of venom also contain cyanide, so, for a species that doesn't have public service announcements and health class, there is a real danger of the monkeys dying from exposure to the toxins.

That could explain why they often get high in groups, although it's not like they have 911, or EMTs standing by. And because capuchin monkeys have absolutely no facility for figurative ideation, when they get together to get high and shoot the shit, they literally get high and throw shit at each other. Which works in their world—and during spring break at Lake Havasu.

WALLABIES

What do we know about wallabies? Well, they are marsupials that live in Australia, and they resemble kangaroos, only smaller. They lend their name to a desert boot–style comfy shoe favored by English professors and NPR listeners. Some wallabies can survive without water, subsisting on the juices they find in plants. And when wallabies cut loose, *they like to chase the dragon.*

That's right—wallabies have been documented doing a *Wizard of Oz* and running through poppy fields. But instead of falling under the spell of the Wicked Witch of the West, these flat-nosed hopheads bend over and chow down on the poppies. After that, it's all sunshine, lollipops, and rainbows everywhere.

And they don't stop there. Like they're in a crazy mash-up of *Trainspotting* meets *Signs*, the wallabies start to get silly and trample on the poppy field. Only they don't just make a mess, they make *crop circles.*

This fact throws me straight into full-on *In Search Of . . .* mode. Were the ancient astronauts that started the Toltec civilization actually highly advanced wallabies just back from a toot at the opium farm? Could wallabies be weaponized for the war on drugs and air-dropped into Colombia and Afghanistan? Could we do a remake of *Basketball Diaries* with an all-wallaby cast? After all, they are addicted to heroin, and those suckers can *jump.*

ELEPHANTS

"Hello, my name is Jumbo, and I'm an alcoholic." These are the words the residents of the eastern Indian town of Dumurkota wish were being spoken in the basement multiuse room of their local ashram. But alas, elephants can't talk, don't use folding chairs, and would never finish all twelve steps.

The only step they took was to raid a Dumurkota shop selling mahua, an alcoholic beverage made from the mahua tree, which is very fragrant. This concoction attracted about fifty elephants to town in 2012. After polishing off nearly 150 gallons of the stuff, the elephants revealed themselves to be not just drunks, but *mean drunks.* Like Icelanders on a weeknight, once they had run through the whole mahua supply at the shop, they began smashing through town, knocking down buildings in search of more of that demon hooch.

Before they could get to that part of the night when they would get all weepy and start talking shit like "My boss is a dick" or "I'm going to call my mother," the villagers managed to drive the elephants away across a river. (Only the most experienced villagers were allowed to do this, and were thus designated drivers.)

There are several other reports of elephants getting into Daddy's liquor cabinet. What could have motivated this behavior? My suspicion is that after millennia of having the best memories in the animal kingdom, these elephants were just drinking to forget.

BEES

What's more terrifying than 50 drunken elephants? How about 50,000 drunken bees?

Fermentation is happening all the time. That white powder on fresh grapes? Yeast. The yeast eats the sugar and poops out alcohol. And sometimes this can happen in nectar, in flowers on the ground, and in the trees.

So, what happens when a bee drinks some spiked nectar and gets hammered? It's not all that different from what happens to humans. According to several sources, it impairs their driving (or in this case, flying) skills, and they get into many more flying-related accidents. Some drunk bees forget how to get back to the hive, which is the equivalent of having one too many at the game and forgetting where you parked, although when you sober up, you're not left on Level G to die. And, like humans, the bees back at home don't take

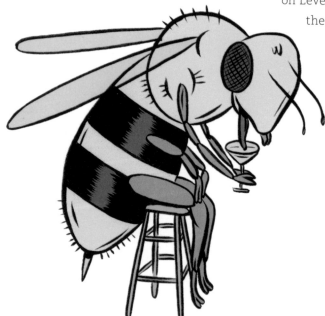

kindly to you showing up with a load on, although humans don't (usually) show their anger and disappointment by *chewing off the drunk's legs*. But bees do! Which is another reason why being a bee is just awful.

In a recent study at the University of Haifa (Go . . .? Damn, they don't have a mascot), researchers discovered that bees prefer nectar with naturally occurring caffeine and nicotine. So, like people, if the bees can't get their drink on, they can at least settle for coffee and a cigarette!

JAGUARS

You know what jaguars are doing in the mysterious jungles of Peru? They are getting high on their own personal ayahuasca retreat, and you're not invited.

Ayahuasca is a hallucinogenic tea that has gained popularity in the United States over the past few decades (though it is illegal). Some accounts say it has effects similar to those of peyote and thus is used by shamans in the Amazon for religious rites and the search for enlightenment. The drug is made from a few sources, among them the *Banisteriopsis caapi* vine—the same vine jaguars eat when they need to void their bowels.

Jaguars regularly purge their digestive system by eating this vine, which not only causes vomiting and diarrhea (in both cats and humans, yet another peyote parallel) but also seems to keep the growth of intestinal parasites in check. The side effects are extreme ball trippage. There are several videos online that show jaguars eating the leaves of the vine, climbing a tree, and watching the universe come pouring into their eyeballs like psychedelic soup.

And by the way, England, it's "jag-wars." Not "jag-you-wars." So, let's stop that nonsense right now.

BIGHORN SHEEP

I once went on a shoot with Tom Green in the Canadian Rockies, and at one preposterously-Canadian-Rocky-style event, our way was impeded by a group of bighorn sheep running all over the damned place. Here's a hint: if you're trying to avoid bighorn sheep, don't climb a hill to do it. Those things *live* for hills, it turns out. And they also live for tripping bighorn balls, by way of munching on some rare, blue-green narcotic lichen that takes decades to grow. Grow naturally, that is—the sheep aren't climbing those hills to get to a hydroponic farming store. Would that it were so, because if they want to get their rocks off they have to get it off rocks, going down on those rocks like a crack addict going down for a rock. They will bloody their gums getting every last bit of lichen off, so they can get off.

VIENNESE SONGBIRDS

While most of this section is about animals that like to go on benders, occasionally nature will slip you a mickey, and that can have disastrous consequences. In Vienna, which gives us the word *wiener* so it is intrinsically one of the funniest places on Earth, it is not uncommon for some species of songbirds to accidentally ingest naturally fermented berries. And not only does this get them high on a hill with a lonely goatherd, it often results in extreme liver damage, injuries, and death. In 2006, Austrian scientists collected approximately 40 songbirds in Vienna that went and got their berries on, then started flying into stuff like windows, trees, and the ground before either breaking their necks or dropping dead from alcohol poisoning. I can only assume that—as the city was full of drunken songbirds—some Wieners can claim to have heard the crunk bird sing.

REINDEER

Listen, I mentioned this in chapter 6, but it just can't be emphasized enough: Reindeer like to eat hallucinogenic mushrooms and trip balls. Then their Laplander deerherds drink their pee and also trip balls. Then they save their pee and drink it again and trip balls. And you say there's nothing to do in Lapland?

Chapter 21

NUMBERS DON'T LIE

Who doesn't love an unusual statistic? Politicians and religious leaders aren't the only ones who can twist the facts to suit their needs: Statisticians like to pile on the data, whether it is relevant or not, to give us some truly mind-boggling tidbits to share with friends and strangers when we have absolutely nothing else to talk about.

One of my favorite ways to look at data is to unfairly compile one statistic, then absurdly contextualize it with something completely unrelated. For example, if you took all the beer Americans consumed in 2014 and put it all together, not only would you have millions of furious and sober Americans, you would also be able to fill Wembley Stadium five times over, with some spillage for the foam. That would be a very stupid thing to do! Or, to paraphrase Joan Rivers's famous joke, if you laid all the women of Wellesley College end to end, it might take you the whole weekend.

Some statistics manage to still be surprising, even though they are based on strict scientific investigation. Like the fact that after doing a visual survey of parked cars in five cities in the United Kingdom in 2012, a team of researchers determined that red cars receive the most bird poop. There's a lot to be surprised about with this: The scientists didn't trust the birds of just one city to give them accurate results, they thought that taking this line of inquiry would somehow benefit humankind, and they did not delve into the cause: Is there some evolutionary mechanism that compels a bird to poop on red objects, or do they just resent men having midlife crises?

You don't use stats just to figure out what's exceptional; often, they are used to find the fat part of the bell curve and tell us what's normal. You might be up all night worrying that you drink too many 5-hour Energy drinks until you find out that in many urban areas, up to one out of three people under thirty consume at least one unregulated, highly caffeinated, high-fructose-corn-syrup beverage a day. So, stop worrying about that and start worrying about the side effects of the Ambien you're taking to counteract that 5-hour Energy drink.

Statistics are great at predicting trends, except when they're not. Look at scholar Thomas Robert Malthus, who authored *An Essay on the Principle of Population* in 1798 and got a whole generation worked up over the idea that if the population kept increasing and food production stayed the same, there wouldn't be enough food for everyone, riots and upheaval would rule the day, and, you know, dystopia ensues. He was wrong about the cause of that last bit, because rather than food-supply problems, we have a pretty good shot at dystopia based on, say, real-estate-mogul-cum-president issues. So, start gathering statistics on canned food and shotgun shells!

Another popular statistical game replaces one object or idea with another. Like, if every molecule in a spoonful of water were blown up to the size of a spoonful of water, that new volume would be greater than all the oceans of the world combined. (Less of a cocktail party factoid and more of a just-got-high-in-college-for-the-first-time wow-factor factoid, but in this new world of marijuana use, one must learn to be tolerant.)

To that end, please be tolerant of the following statistics. And if you don't find them mind-blowing, maybe you should, you know, "experiment." According to the CRC Health Group, the average first-time marijuana user is 17.0 years old, so put that in your pipe and smoke it.

PUTTING IT TOGETHER

People love to look at statistics about the excesses in American behavior and spending. How fat we are. How much television we watch. But food is awesome and TV has never been better, so shut your faces!

Perhaps the most unfair statistics that show up from time to time concern how much we spend on our pets, because, when you add it all up and stick it in an article, we spend an astonishing amount. Sixty-three billion dollars a year, just on our pets. That's more than the gross domestic product of Uzbekistan, a country known for its extremely gross domestic products. But don't be alarmed, because study after study has indicated that pet owners, particularly dog owners, enjoy greater health benefits from having the companionship and responsibility of having a pet around. About 20 percent more households have dogs versus households with cats, and while there is overlap, cat owners are twice as likely to have more than one cat than dog owners are to have more than one dog. So, the crazy-cat-lady theory holds up.

Dog owners are always running around claiming that dogs are better than cats, because they are. But here's a fun statistic to chew on: Between 2004 and 2015 in the United States, there were ten cases of dogs shooting their owners, while only one case involved a cat. To be fair, all the dogs were hunting dogs and each incident was a matter of mishandling a weapon—don't go eating Slim Jims out in the woods and expect your dog to leave your greasy trigger alone—but in the case of the cat, it is a foregone conclusion that the perp meant to do it, because cats are awful and dogs are not, though I have no statistical evidence to support that.

ODDITIES AND SURPRISES

While I was writing for Jay Leno on *The Tonight Show*, we developed a species of joke we called "Duh Science." These dealt with studies that proved stuff that was already obvious to everyone. For example, the University of Alabama at Birmingham took their shot at scientific immortality by conducting a study that proved that walking in high heels makes your feet hurt. The American Heart Association got back to basics with a 2014 study that shows quitting smoking improves your quality of life. And in 2016, the University of Cambridge published the results of a study proving that Spider-Man could not exist. I've been saying Cambridge was stupid for years—maybe now people will listen.

We never made an *Edumacation* segment out of studies that were surprising, because, by their very nature, they are few and far between. But there are some cool ones.

I'm intrigued by the kinda scary ones. Like the one that concluded that of all the days of the year, you are most likely to die on your birthday. Go, Andy! It's your birthday! And when we say, "Go," we really mean *go*! Or, as an American male, I am more likely to die from breast cancer than testicular cancer (but much less likely to get it in the first place). Or, were I to get a woman pregnant, I would be more likely to father twins than I would be to win the Pick 3 in the state lottery. Which puts this in the scary paragraph.

My favorite surprising statistic is something you will cover in your Statistics 101 class in college. It's called the Monty Hall problem, because it has to do with picking the door that has the prize behind it on the old *Let's Make a Deal* game show, hosted by program cocreator Monty Hall.

It works like this: You dress in a humiliating outfit and carry all sorts of crap in your pockets so Monty will pick you, and you make some smaller deals before you get to the big-game level. Then he shows you three doors, behind which are two donkeys and a car. He asks you to pick one. Simple, right? At this point, you have a one-in-three chance of getting the car. But then, before you find out whether you win, he *opens one of the other two doors, revealing one of the donkeys.* Then he asks you if you want to change your answer to another door. Sounds trivial, right? I mean, what's the difference between one door and another when they're all still closed? But no—if you *switch* your answer, you have *twice* the likelihood of getting the prize.

Why? Because you start out with a one-in-three chance. If you hold, your chances stay the same. And since the open door is not the winner, the closed door you didn't pick has the rest of the chances, which comes to two out of three! Time to go to YouTube and watch all the assholes who didn't switch. Assholes!

NORM!

A lot of stuff considered normal to a statistician is often surprising to the rest of us. Just look for any article that begins with "The average American home . . ." and you'll see what I mean.

To wit: The average American home has 300,000 objects in it. If you were to try to count all those objects out loud, it would take you eighty-three hours, assuming you didn't sleep but still counted the numerous cups of coffee you would need to prove something as pointless as that. But remember, it includes all the change you throw into that cup by your bedside, so maybe if you finally went through all your shit, something good would come of it.

And while we're on the subject of how much crap you have, you have a lot of crap—so much so that one in ten Americans rents offsite storage, and they rent a shitload of it. There are five times as many private storage facilities in the United States as there are Starbucks coffeehouses. And there are 7.3 square feet of storage for every man, woman, and child in our country, so if the Decepticons finally show up, all we have to do is run for the nearest Stor-All and we'll be covered. And thank God for those storage units—even with them, one-fourth of Americans with two-car garages can't park in them because of all their shit, and one-third can get only one car in. How do I know so much about personal storage? Because I want to rent a unit, and I want my wife to know it's perfectly normal.

US homes have more television sets than people. Americans spend a hundred billion dollars a year on shoes, jewelry, and watches, which is more than we spend on higher education. The average woman spends more than eight years shopping. And in the United Kingdom, a recent study showed that the average ten-year-old owns 238 toys but plays with only 12 daily. I threw that in there because it's meant to surprise me, but really, how many toys can a kid play with a day—especially a British kid who is powered by Marmite and has to use all those extra "u's" while writing?

TRENDS AND PREDICTIONS

Before we had Moore's law, which states that processing power in computer chips and storage will double every two years, there were a ton of predictions out there about the future of computers. You have to realize that compared with what we have now, even doubling that capacity can feel like an endgame to an armchair prognosticator. My favorite prediction comes from *Popular Mechanics* in 1949, in an article about the (then) amazing ENIAC computer. "Where a calculator like ENIAC today is equipped with 18,000 vacuum tubes and weighs 30 tons, computers in the future may have only 1,000 vacuum tubes and perhaps weigh only 1.5 tons." Progress! Remember, this was a time when computer bugs were actual insects, so give them a break.

In 1966, while working on the *2001: A Space Odyssey* book and movie with Stanley Kubrick, Arthur C. Clarke was asked for his vision of 2001, for obvious reasons. One thing stuck out to him: that our homes would be so self-sufficient, that recycling would be so perfect, that power sources would be so small and efficient, that entire communities would be able to migrate south for the winter to follow the good weather. I guess when he moved to Sri Lanka, he completely missed out on the idea of Boca Raton, and of course this prediction did not come true, unless you count the trailer people in *What's Eating Gilbert Grape*.

Nutrition, general health, and exercise have caused the human species to get taller over the years. This was underlined in a 2004 study of Bayaka pygmies of central Africa, who turned out to be short due to malnourishment. But scientists in the first half of the twentieth century didn't quite grasp the whole idea of genetic potential, so based on average height data from the 1850s to the 1950s, they predicted that the average woman would be 6-foot-3 by 2000. That's nearly *two* Kristin Chenoweths!

IDEA SWAP

There's been a lot of talk about the national debt and the trade deficit lately, and while most of us know the difference, it seems like the new guys in Washington don't. But we will put that aside for now and go for a classic in the Idea Swap category: If you were to take the national debt and convert it to dollar bills, then stack them up, how tall would that stack be? As of this writing, the national debt is $18,152,809,942,589, which ain't hay. So, how tall would 18 trillion bucks be? According to an article in *Wired* magazine, a dollar bill is 0.1 millimeter thick, so 10 to a millimeter, 100 to a centimeter, 10,000 to a meter, and so on. Seems like we're running out of money quick, right? NO!

A trillion dollars would get us about a quarter of the way to the moon. So the national debt stack would get us to the moon and back twice with plenty of pocket change. Which is actually pretty close to the cost of doing two full moon missions, so maybe we should just stack the money up and let the astronauts climb it.

Finally, according to a theory proposed by John Wheeler to Richard Feynman in the spring of 1940, if you had a dollar for every electron in the universe, now and for all time in the past and future, how much money would you have? A dollar. Because his theory is that there is only one electron, ever. Trick question!

Chapter 22

DANGEROUSLY EXOTIC VACATIONS

A broad title, and it doesn't cover everything that could kill you on vacation. For example, say you're a dad. You might be traveling with the wife and kids, and to save a few bucks, you decide to get a couple of queen-size beds and maybe a cot, so you can all stay in the same room together. Then you snore all night, approximating the sound and volume of someone slapping a 4-pound slice of beef against the side of a bathtub. Thinking only of their own sanity, your family drowns out the sound by covering your face with a pillow and, for good measure, pounding it into your face with a heavy lamp. Even though this happened to me on two occasions, we will not be covering it in this chapter.

Instead, we will look at the creatures of the natural world that can make life a little shorter for you. It is a common practice for city dwellers like myself to take their families somewhere less citified. We're not talking about a jungle safari—maybe a stay at a beach club on a nice island. Maybe we'll take a nature walk near a hotel. Or even do a little snorkeling to have a look at the reefs before they're all dead.

Yet in each of these places, there is probably something out there that can kill you—and for some reason, you will probably take chances abroad that you would not take at home.

For example, if a couple of grinning yahoos wearing shorts and wraparound sunglasses were to ask you if you want to zip-line between buildings in Camden, New Jersey, you would probably try to find a clever way to tell them to go climb their thumbs. Yet south of the border in Mexico, where decapitation (intentional or otherwise) is as common as athlete's foot, you are perfectly willing to fork over eighty bucks for the privilege of experiencing this "adventure."

It's especially true with critters. Tourists view regional animals as "exotic" and are delighted to see them in their natural habitat. But the key concept is that they're not in a zoo, with carefully planned fences and trenches between you and Mother Nature.

And we're not even talking about the critters that, if you come in contact with them, you wish *would* kill you. Friends from Texas went to a beach resort in Micronesia, very exclusive, which means you have only about ten other couples you have to get along with, and there's no doctor. The wife wasn't feeling well, but she started to get over it, so they lay out on the beach. After about an hour, a 9-inch parasite wriggled out of her butt. The staff caught it and put it in a Ziploc bag so the doctor could have a look when she got home. Believe it or not, they left early. ("Do you have anything to declare?" "Besides the intestinal parasite that's kicking up a fuss in my suitcase? No.") Would you want to live after giving butt birth to that? Not me.

One quick aside: This chapter is organized by location, but all across the world there are wonderfully stupid ways to die on vacation. Scalding yourself in a hot bath? Popular in Japan. Driving on the wrong side of the road? Ain't nothing commoner to a colonial commoner. Getting killed while taking a selfie? Way too frequent—and, by the way, you deserve it (says the man who bought his selfie stick at an airport for $116).

Intestinal flukes aside, there are creatures around the world that can send you home in the luggage hold. This is by no means a comprehensive list—there are plenty of other ways to kick the hotel ice bucket. After all, it's a small room, and everyone in your family has a pillow.

HAWAII

The Aloha State is truly a paradise. Since we have very few paradises here on Earth, we tend to treat it with hyperbole, touting the good stuff like great weather, nice people, and some of the oldest marijuana farms in the United States. But beneath that garland-giving, surf-riding, lava-spewing wonder lie creatures intent on sending you off to your metaphorical paradise.

First, stuff you should know likely won't kill you: Shark attacks are rare in Hawaii, so you can relax about that. Waikiki Beach has incredibly easy-to-ride (if small) surf because the reef is so shallow and extends so far into the Pacific. But you still see sharks, which can be upsetting for a human. Also, mercury poisoning is not uncommon if you eat a lot of fish, because it's a volcanic island chain with a lot of venting mercury and the fish you eat in restaurants tend to be larger specimens that accumulate lots of mercury over time. It won't kill you, though, because it's treatable.

But then there's the yellow-bellied sea snake—which has a supremely appropriate name, since not only is it a sea snake, it also has a yellow belly (and is very timid). You know, like in every episode of every western TV show, when someone is a coward, they are accused of being yellow bellied. And this snake is both! It is one of the most poisonous marine reptiles, but thanks to its cowardly streak, it is very rare for anyone to even see one, let alone get bitten by one. But if you're one of those people who just has to slap a snake whenever you see one, you may want to avoid the Hawaiian Islands, just in case.

However, more people are killed in traffic accidents than suffer any other untimely death in Hawaii, so if you find yourself between a yellow-bellied sea snake and rush hour traffic on H-1, go ahead and slap that snake.

COSTA RICA

Actor, comedian, and Canadian Tom Green was a correspondent for *The Tonight Show*, and I was his writer. One day, he decided to buy a beachfront house in Costa Rica, because he is prone to making irrational decisions. Tom puts the "rash" in *irrational*. Since the house abuts a nature preserve that is home to two species of poo-throwing monkeys, Tom wanted to shoot a comedy travel piece there. But NBC (for insurance purposes, or maybe just to fuck with me) made me take antimalarial drugs, as well as get shot up with drugs that would protect me from everything from rusty nails to poop-borne grossness. The only problem? The antimalarial drug caused me to hallucinate extra doors in my office and made me want to kill myself.

Luckily, Tom went on the trip early and was dashed against the rocks by a rogue wave, breaking two ribs and having to endure a six-hour drive on a pitted jungle road to get medical attention, which canceled the bit. Problem solved!

If we had gone, the crew and I might not have made it back. Based on its animal population, Costa Rica takes the title for most dangerous in the Americas.

Let's say you decide to travel there and go in the water as a human with something to contribute, maybe even something to say; you could come out as another animal's poop. Costa Rica is responsible for the most deaths by crocodile in our hemisphere, which is an awful way to go. Even if you get only partially converted to poop, the teeth marks on the remaining parts are definitely going to get you a lot of left swipes on Tinder.

So, maybe you stay out of the water and find a nice house on 2 acres of land—which means you have a one-in-two chance of encountering a deadly snake. That wasn't in the brochure! The most dangerous snake is not the most deadly, but it is such an aggressive jerk, it's the one you'll probably get bitten by. It's called the fer-de-lance, or spearhead, and it puts about 600 people in the hospital per year.

If you met a lot of nice people in the overcrowded snake ward, maybe you want to give it one more go. You walk out of the hospital and head for the rental car when you feel a tiny bite on your neck. And that mosquito—the world-champion killer responsible for more deaths than all other animal attacks combined—is about to suck your blood. But before that, she gets just a few viruses into the wound she made, and they go to work. Dengue fever? A few years ago, 2013 was a record year, with 40,000 cases. Zika? Costa Rica freakin' *invented* Zika. And of course there's malaria. In Costa Rica, there's *always* malaria.

But on the other hand, the food is delicious!

UNITED KINGDOM

Despite there being plenty of British murders in imported television dramas, when it comes to the animal kingdom, jolly old England seems pretty darn safe. In spite of the fact that all their food is identical to ours but packaged in weird, unsettling ways (the same can be said of their language), nonhuman creatures there are basically pussies, right? Not so fast, you ugly American, you!

If you bear any resemblance to Tippi Hedren, first of all, congratulations, but second, stay away from the British Isles. Because there is a growing threat of *seagull attacks*. That's right, these Jonathan Livingston Assholes have taken to attacking anyone with the balls to eat food of any kind while outdoors and near the coast. In a typical onslaught, they come in low and strafe you with vomit and poop. If you run but don't drop the food, they come in fast and claw at you, and they have caused large, lacerative wounds on people's scalps—all because you wouldn't give up your French fries (or chips, as the Brits call them, which makes no sense).

But at least they don't kill you. Bandaged and hungry, you make your way inland to get away from it all and explore the pastoral bliss of English farmland. Rolling hills, ancient hedgerows, and cows. Lovely cows, many of which are named for the very land they graze. It's all so beautiful. Look, there's a baby calf! You take out your smartphone. You climb the fence—and then this one cow hauls ass, runs you down, and stomps you to death. Because cows are the deadliest large animals in the United Kingdom, responsible for killing an average of two people a year. You, my friend, just got killed by a cow. Let *that* sink in.

But say that by some miracle, you survive both the attack and the Nationalized Health Service's cow-recovery wing at Whatevershire Hospital. You're walking out of the hospital toward your rental car when you feel a tiny sting on the back of your neck and, almost immediately, things go south. Bee and wasp stings account for the most deaths by animal attack in the United Kingdom each year, not so much from the sting but from going into shock from the venom. On average, five people a year die in the United Kingdom from bee and wasp stings, making it more likely that you will die that way than from a terrorist attack.

And as you lay dying in that miserable gray parking lot, your ears ringing impossibly loud, your hands and feet and lips swelling up, your throat closing, you have time for one last thought:

"At least I'm not in Costa Rica!"

AUSTRALIA

Now we're talking. Australia is not a country where you *might* get killed on vacation, it is a country that is going to go out of its way to kill you in every conceivable sense of the word. In Hawaii, they give a garland necklace when you get there. Australia should give you a suit of armor, an elephant gun, and an EpiPen.

First up, this is a place where the sharks *will* kill you: They kill three people a year, on average. And before you say, "So what?" that's only two less than the number of shark attacks in *Jaws*! (We are going to be using a lot of exclamation points here; we are in Australia, after all.)

Worried about deadly snakes? How about *the deadliest snake in the world?* The inland taipan has not been responsible for any recorded deaths, but it has enough venom in one bite to kill several people. And why does it need to be so deadly, you ask? Because its prey is the scarily named plague rat, *one of the world's toughest rodents, which can kill most other snakes!* Because, besides the opera house, Australia is a miserable prehistoric hellhole!

It would make a lot of sense for Australian officials to at least issue flamethrowers to all visitors as they arrive at the airport, if for no other reason than that Australia has a particularly scary spider. And not just any spider, but (say it with me) *the deadliest spider in the world!* The funnel web spider's venom is uniquely poisonous to humans, though no one has died of a bite since 1981, when Australia developed an antivenin (and finally got running water).

They also have not one but the top *two* deadliest jellyfish, the aggressive bull shark, lethal brown snakes, their own flavor of crocodile, cone shells, the blue-ringed octopus . . . the list goes on and on. They even have the common death adder—and I can't tell which is more common, the adder or the death.

With apologies to Dr. Karl, a popular TV and radio science whiz Down Under, Australia is the most terrifying place on Earth! Though, as I said, that opera house is pretty sweet.

Chapter 23

SAY WHAT?

When I was assigned to take Tom Green to the 2006 Winter Olympics in Sestriere-Colle in the Italian Alps, my idea of what was going to happen was as far from what actually happened as you could imagine.

What I expected from the Alps was that they'd be impressive and huge, just one Matterhorn after the other poking holes in the upper atmosphere. Centuries-old taverns and inns would be side by side with sleek, modern chalets and state-of-the-art skiing facilities. And the Olympic crowds would converge from all over the world to watch the finest in winter sports while our Italian hosts—in the country that taught the world to eat, and pray, and love (but not in that order)—would roll out the red carpet.

What I got from the Alps was essentially the Mount Airy Lodge in the Poconos, with two pizza places and a bar. The stands were empty, and the carabinieri—

the national police of Italy, named for the machine guns each of them carries—hated everything and would walk off the job all the time. Our edit bay was in a temporary trailer with a crack in the floor, which was freezing until we realized we could seal the crack with Pepsi, because it was 10 below at night and none of us drank Pepsi. We all got food poisoning so consistently from our hotel restaurant that they should have listed it as a side dish on the menu. As far as the production went . . .

Tom's drink of choice was ouzo, so everywhere Tom went smelled like a Good & Plenty factory. During one taping, he stole a gypsy's accordion and I was threatened by a nine-year-old with a knife. But the one bar in town was owned by a Tom Green fan who gave Tom his own table every night, where he was joined for hours by champion skier Bodie Miller. Of course, on this trip, Bodie Miller hit five gates one morning and had the worst showing of any skier in that Olympics. As for me, I broke a tooth because I had been vomiting so much from the food poisoning that it weakened the enamel, and I lost 36 pounds in three and a half weeks. My cameraman snuck off and broke his arm snowboarding with five days left to tape. And everyone back at the office hated my guts because I got to go to Italy.

I was being sent away from my family for essentially a month, and on top of that, NBC forced all of us to take Italian lessons so that we would be the goodwill ambassadors in NBC production jackets. The irony was that from the first moment we arrived until almost the last second of taping, *everybody spoke English*. It was completely *pazzo*.

But then, at the very end, during the live throw from Torino, I was standing by the barricades and an old man tugged on my sleeve. In careful Italian, he explained that he lived there, had never been farther than 30 miles from his house (where he was born), and was grateful that we came and put on a show. And I understood him! I felt better all of a sudden, and in my horrible Italian (which I've since forgotten) I said it wasn't easy. He asked me if I had *una brutta vacanza*, and I said, "Si! Si!" And he rolled his eyes and walked off.

I later learned he was mocking me for having a bad vacation. He was just like those assholes back at the office!

But I did pick up some weird phrases from Italy, and added them to my collection of odd ways of saying things from across the globe.

I would like to put in a disclaimer here: The author of this book and its editors and publishers are not responsible for broken noses, lost teeth, and cracked sternums that result from readers trying out these expressions in their native countries.

EUROPE

Europeans migrated out of Africa tens of thousands of years ago, but if you look at a map of both continents, you'll know the Europeans are even more tribal than their southern forebears. Sure, the decrease in sunlight genetically selected for generations with lighter skin and the ability to digest milk, but what it really seems to have done is make them spend those long winter nights coming up with stupid ways to say stuff.

It's no surprise that in Italy, a lot of their expressions come from food. Say you break up with your partner, but a lifetime of watching sitcoms convinces you to try to put that unworkable relationship back together. That would be called *cavoli riscaldati*, or "reheated cabbage." And while the English say, "I'll make mincemeat out of you" (especially if they're Klondike Kat from *Underdog*), the Italians have their own threatening meat-grinding phrase of choice, *fare polpette di calcuno,* or "I'll turn you into a meatball."

The Germans like their meats, but they also have that violent streak you might have noticed if you've ever taken a history class or watched an *Indiana Jones* movie. Ever experience gaining breakup weight? Done some stress eating, and now you can pinch more than an inch? The Germans call this new brie jumpsuit you're wearing under your skin *kummerspeck,* literally translated "grief bacon." (No better way to grieve.) But say your ex decides to be a dick about it, calling you out for not fitting into your Elvis Costello North American Tour T-shirt. You might say they have a *backpfeifengesicht,* or a "face badly in need of a fist." How much do the Germans save by not using spaces in their words?

The French have a great expression, *l'espirit de l'escalier,* or "stairway wit," which refers to coming up with the most awesome comeback long after the situation is over. And in that spirit, "Oh, yeah? Well at least I *have* a pencil!" (That was for Tom Thornton in eleventh grade. He knows what I'm talking about.)

And remember, just because you hear a familiar word in another language doesn't mean it has the same meaning. In Norway, the funniest word in all of history—*fart*—means "speed." So, you can get pulled over for farting. Keanu Reeves saves Sandra Bullock in *Fart*. (Nobody saw that awful *Fart II*.) *Top Gun* would have been infinitely better if Tom Cruise had said, "I feel the need—the need for *fart*." And I'll never be able to watch *Star Wars* without watching Chewie make the calculations to jump to light*fart*.

At this point, I would like to thank my publisher for that paragraph. I have a very immature bucket list.

ASIA

The Philippines, for all its craziness, has some pretty handy expressions that I hope to incorporate into my daily speech and then explain to people why I'm speaking Tagalog. Oddly, they have no word for "driving on a Vespa with my entire family, two dogs, and a 60-gallon fish tank," but they do have the word *layogenic*. No, it is not a perverted form of *photogenic*—it actually means "cute from afar, but far from cute." After shooting several segments for Jay Leno with *The Sopranos*'s Steve Schirripa (who is just awesome, by the way), I discovered that the proper North Jersey term for this is *butterface*." Don't shoot the messenger.

Indonesia just wants to confuse the hell out of English speakers, so while the following examples are not technically odd phrases, they're odd to me: *Air* means "water." (That is a black-is-white, up-is-down situation right there.) And you know that stupid middle school trick of going up to someone, then tapping them from behind on the opposite shoulder so they turn to look away from you? They have a word for that—*mencolek*. That is nothing short of straight-up fucking with you. What's amazing is they don't have a word for pointing at your chest and, when you look down, *booping* your face—or, indeed, a shorthand for the classic "Got your nose." But let's give them time, people.

The Japanese are famously polite. After college, my dear friend Susan went to work at a big American bank in Tokyo, and she told me later that on the subway, countless men would give her impromptu breast exams entirely free of charge. (Do you know how much 26,000 breast exams would cost in today's healthcare crisis? Of course you don't—nobody does.) Another cool and super-male thing in Japan is the word *bakku-shan*, which describes a woman who is attractive from behind but not the front. (See Steve Schirripa reference above.) And say you love slipping into a 1,000-yard stare, but you're on a subway and all you can see is an attractive woman from behind getting a free breast exam. *Boketto* means "blank stare" in Japan.

BRITISH ISLES

There are plenty of British phrases I think we're all used to, thanks to Monty Python and Harry Potter: "Bob's your uncle," *bollocks*, *chuffed*, *cock-up*, "get stuffed." George Bernard Shaw called the United Kingdom and the United States "two countries separated by a common language," and it don't get no commoner than what I'm about to discuss.

Say you're in Ireland, a country where history and stuff happened, but, more importantly, is a filming location for *Game of Thrones*. You try to establish a rapport with the locals, but cries of "Winter is coming!" are met only with nods of agreement, since you are in Ireland and winter pretty much comes every year. But then things get really confusing when you ask about Sean Bean, and they bring out an old woman. After a lot of confusing and angry exchanges, you realize that *seanbhean* is the Irish Gaelic word for "old woman." Which explains the haircut in *Game of Thrones*, am I right?

Having avoided a fight and met a nice old lady, you decide to stick with the *Game of Thrones* theme, which you think pays off when someone mentions that you have a crusty dragon on your face. Fear gives way to embarrassment, however, when you realize they mean you have a booger, then embarrassment gives way to sadness when you realize you will never have sex with Khaleesi.

And don't say *fanny* in England. Here, it is a semipolite way of referring to a person's posterior, but over in Blighty, it means the lower front in a lady's ladyparts. For a country that uses the C-word so liberally, this seems odd. It also explains a lot of the female bonding in Fanny Flagg's *Fried Green Tomatoes at the Whistle Stop Café*. So, whatever you do, don't say, "Fanny!"

What can be very confusing for Americans in the United Kingdom is the astonishing number of terms the British use to describe a penis. Which I will tell you now: *bell-end*, *cock*, "crown jewels," *dick*, *dong*, "giggle stick," "John Thomas," *johnson*, *joystick*, *knob*, "meat puppet," *parsnip*, *pecker*, *pizzle*, *plonker*, "pocket rocket," "pork sword," "purple parsnip," "swinging cod," *tallywacker*, "thirsty ferret," *todger*, "trouser snake," "wedding tackle," *willie*, and *winkle*. That is only a partial list, so if someone in the British Isle hurls an epithet at you that you don't quite get, you are probably being called a dick.

THE AMERICAS

Latin America. Lands of mystery to many Americans since, to paraphrase Steve Martin, "They've got a different word for *everything*." And even if you understand the individual words, you may not understand what the hell they're saying, because they use some pretty damned odd expressions.

Venezuela: They know how to raise and age fine steak, and they know how to turn a phrase. To raise that beef, you have to be practical, hard-working, efficient, attentive. So, for those of you who would rather build castles in the air, daydreaming your life away, they have an expression for you: *"vevir en nube de pedos."* Which translates to "living on a cloud of farts." Cows are the biggest methane producers in the world. Coincidence?

Argentina also has wonderful beef. But if you wait too long to feed some to an Argentinian, they might get hungry. Their stomachs might start growling. They might exclaim, *"Me pica el bagre!"* Which means, "The catfish is biting me!" Because every man, woman, and child in Argentina has a catfish living in their intestines. Fun fact.

Just like in the American South, residents of Latin America have a deep and justified distrust of people from the north. Another similarity is how they refer to soft drinks: If someone were to offer you a soda, be it a Fanta, a Faygo, or a Squirt, they would ask, *"Que tipo de coca-cola quieres?"* That's right—just like south Georgians, they refer to all sodas as Coke. However, they do not refer to all chemical stimulants as cocaine, which would make a lot of sense (to me, at least).

Apparently, all across Latin America, they might have capybaras, chinchillas, nutria—but they don't have shithouse rats. (Perhaps it is due to the catfish that live in their intestines.) So, when they want to make a comparison between the mental state of an insane person and an animal in an unusual living situation, they will say *"mas loco que un cabra con pollitos!"*—or "crazier than a goat with chicks."

We need to get those guys some shithouse rats!

Chapter 24

THE SCIENCE OF WEED

What a time in the book to get serious.

But I'm going to have to do it, because new discoveries about marijuana—both its long-term medical effects and its use in treatments for ailments from arthritis to epilepsy—are finally coming to light, and it may turn out that marijuana may be one of the most important drugs in use today. Before you ask me if I'm high, the answer is no, because I am one of those people who becomes almost completely nonfunctional on the stuff. To wit:

Kevin Smith is a proponent of marijuana use. I work with Kevin Smith. Kevin has always wanted to get me to smoke with him, since he thinks it could help me with my anxiety and assist my creative energies. So, one day I said yes. We picked a milestone—the final episode of *The Tonight Show,* and, after the weird wrap party in the NBC Studios parking lot, I went over to Kevin's house to record a new episode of the *Edumacation* podcast. I didn't want to drive after smoking, so we agreed that I'd come over and do the usual round of light edits on Adobe Audition on my MacBook (we used to take out coughs or huge pauses back then, but now we just let it roll).

The next morning, I showed up at nine o'clock and took out my computer. Kevin took out the joint he had packed for me—laughably small next to his, maybe one or two drags in the thing. Being a very cautious person, I decided to just have half a puff and see how things went. Kevin smokes the newest and best hybrids, and when we were doing the *Fox Shortcoms Comedy Hour* pilot at Television City, a famous showrunner came over to hang out with us by the Vespa. He took one hit of Kevin's joint, and you could just hear the dial tone blaring out of his ears.

So I took a tiny hit and put on my headphones, and while Kevin read Google News across the partners' desk, I started to edit the audio using software I have used a hundred times. Then things started to become difficult. I kept taking off my earphones and looking behind me at the door, convinced that someone had entered. I couldn't properly mark the ins and outs of my edits. At one point, I deleted the whole file. Then I got the Mac beach ball.

I never get the beach ball. I am the guy who prides himself on never getting the beach ball. I'm the guy who opens a Bash window and sudos the shit out of that MacBook before he gets the beach ball.

After I spent approximately 350 hours looking at the beach ball, Kevin noticed that I hadn't moved and might have stopped breathing. He gently closed my laptop and told me he was going to get me some food. While he was gone, Shecky (his dachshund) and I had a staring contest, wherein she was either reading my thoughts (pretty much a blank page at that point) or was listening intently to the dial tone blaring from my ears.

After a light snack of garlic knots and burrata from Pizzeria Mozza, my consciousness returned to my body and I got back to editing the show. We didn't speak much of the incident—we just went on working together the way we had been. But for a few weeks, that tiny, barely smoked joint sat on the edge of his ashtray, pointed my way, mocking me.

So, as they say on most warning labels, results may vary.

I have nothing against the stuff. Some of my best friends swear by it. And based on the following, I think it is reasonable to say that weed just might be a miracle drug.

THE STIGMA

When President Nixon signed the Controlled Substances Act in the 1970s, many drugs were classified into different schedules that, on the face of it, were meant to indicate the severity and danger of using them. Schedule I, the most dangerous category, includes heroin, peyote, and LSD—and marijuana. Meanwhile, cocaine and methamphetamine are Schedule II drugs. And so, even though in the 1880s, marijuana grew in most of the empty lots of Manhattan and was found almost everywhere, the punishments for possession, sale, and cultivation were extremely severe.

How come?

It had less to do with how common marijuana was and more to do with how little research was done on it. Even though you could wave twenty bucks anywhere within 3 miles of the Hampshire College campus and end up with a bag of Vermont's finest in the front pocket of your jerga, obtaining medical-grade marijuana was very difficult, and the studies being done were mostly investigations into how marijuana can incapacitate and cause harm.

Meanwhile, methamphetamine was developed to keep World War II pilots awake for long missions, and Norman Ohler's best-selling book *Blitzed: Drugs in the Third Reich* shows how the entire German army and most of the population were riding the snake from the very beginning of the war. Churchill couldn't believe how German troops were making 30 miles a day on foot—it was unprecedented in the history of warfare (because the ancient Romans weren't jacked up on meth).

So, the government knew all about meth, thus Schedule II.

And since this was the 1970s, cocaine was designated a Schedule II drug for fear that, if it was a Schedule I drug, it might take ten years for disco to be developed and the 1980s might never have arrived. (Dan Aykroyd claimed in a 2013 *Vanity Fair* feature that *The Blues Brothers* had a line in the budget for cocaine.) MTV would never have happened. Both Coreys would still be alive. So, Schedule II.

Marijuana was also considered a "gateway drug." Growing up, I thought that was a reference to the Gateway Shopping Center near Valley Forge, which had a triplex cinema that usually smelled like weed. But it actually referred to the unfounded fear that if a kid was willing to try weed, they would eventually go through the entire Haight-Ashbury pharmacopeia and become a useless drain on society.

One thing that has turned out to be true about marijuana fears has to do with users under twenty-five. While the drug is rarely addictive, it can hook heavy users whose brains are still developing. It can also hinder certain kinds of neural growth and reduce adult cognitive abilities in users who start at a young age.

HOW IT WORKS

There was a famous PSA you can still find on the web, directed by reputedly asshole-ish commercial director Joe Pytka, in which an actor holds up an egg and says, "This is your brain." Then he cracks the egg and puts it in a frying pan and starts cooking it sunny side up, saying, "This is your brain on drugs." After a moment, he ends with, "Any questions?" It was the most rhetorical question of the 1980s. While it seemed to do nothing to end the war on drugs (a study in the 1990s showed that antidrug ads actually *increased* awareness and use of illicit drugs in teenagers), it did let diner comedians like myself try to get a laugh from exhausted waitresses by ordering "two brains on drugs, with a side of toast and bacon."

What does that story have to do with how marijuana works? Well, first, it let me get to that diner joke, and second, it shows that fear does nothing to educate and usually causes more harm than good.

Your brain contains billions of receptor sites for different chemicals to not only transmit thoughts and emotions but also trigger different processes in the body and generally get the work of the brain done. Many of these receptors are activated by cannabinoids, which are present in hundreds of varieties in

marijuana. Tetrahydrocannabinol, or THC, is the psychoactive one that produces the effects seen in the introduction to this chapter. But all of them can have some effect, whether in the production of dopamine and other mood-altering compounds (the stuff that gives you the high, a sense of relaxation or euphoria) or in the blocking of these receptors, which can be extremely helpful in some brains that are malfunctioning.

One way that the overall effects, particularly the symptoms you experience (or enjoy, in other people's cases), have been explained to me is that the cannabinoids block many of the so-called hardwired processes of the brain. The more you get into a routine, the more your brain makes dedicated pathways to suit those tasks or processes, or even ways of thinking. Smoking marijuana makes your brain go through the process of starting over, which gives you the feeling of novelty at something you've done (or a movie or a TV show you've seen).

Kevin always claims that smoking weed "shuts out the noise" and lets him concentrate on the task at hand—and to an extent, there's some truth to that. Cannabinoids keep the neurons firing in the brain, which increases our ability to concentrate by inducing a fixating feeling on whatever activity is in front of us. However, that is not predictable, and the shiny-object theory of stonerdom has different effects on different folks. That's why some people are able to drive well under the influence, while others are distracted by thoughts or things they see from behind the wheel.

LONG-TERM EFFECTS

The good news is that most studies show that marijuana does not have any lasting long-term effects on memory or brain function, including concentration. So, if you know a long-term marijuana user and they seem to have impaired memory and can't concentrate, it probably has less to do with them having frequently gotten high and more that they are probably high at the moment.

It's not an instant thing, however. Like most drugs, marijuana has a half-life. Most of the studies indicate that it takes about three weeks after stopping the schmoke before attention and concentration return to normal—and that is from users who used once in the study, or classified themselves as light users. Heavy pot smokers still showed signs of memory and cognitive impairment twenty-eight days after their last ingestion of THC.

The bad news is that if you gotta smoke, instead of taking edibles or vaping, you might get jammed up later on. Two studies indicate that smoking marijuana is much more harmful and potentially cancer causing than smoking cigarettes. However, looking into the methods behind the studies, they examine only the concentration of cancer-causing compounds in raw burned marijuana, and there is no indication as to the quality of the weed, or even if they are comparing it to filtered or low-tar cigarettes. It should be noted that the higher the concentration of THC in the particular strain you are smoking, the less likely you are to smoke dangerous levels of other compounds, because you are literally smoking less. So, that weed your half-friend in college grew under the power lines might actually be more harmful than modern weed, since you have to smoke 4 pounds of it before you can even remotely start enjoying *Fantasia*.

The not-so-bad, not-so-good news is that, yes, marijuana is less addictive and harmful than other drugs. The famous 1999 report from the Institute of Medicine shows not only that that venerated institution has the least creative name of all time but also that marijuana is less addictive—or, to be more accurate, causes a lower percentage of users to become addicted to it—than other common intoxicants. There are so many factors that go into this that it is almost a spurious detail, but here are the results, just the same: Of the number of people *who have ever used* tobacco, 32 percent get hooked. The figure for heroin is 23 percent; for cocaine, 17 percent, for alcohol, 15 percent, and for antianxiety drugs, 9 percent—which are tied for last place with the Mary Jane. But take that first figure with a grain of salt: At the time the study was published, only 2 percent of the US general population had ever tried heroin, and now that number is going through the roof due to the widespread overprescription of opiates and to the super-cheap heroin available nowadays.

In any event, these are the long-term physical effects. Long-term social effects can be damaging since, especially between generations, walking around high on the pot (my mother's expression) can be considered antisocial. And while it doesn't have the same side effects as, say, alcohol, getting high on the job is still messing with an intoxicant on the job and can lead to a step down on the socioeconomic ladder. We had a security guard at NBC Studios who was high all the time, and his solution was to turn his keyboard around at the front gate and have guests type their own names, because he literally had no idea what he was doing. Maybe you've met him if you've ever been to *The Tonight Show* and have a last name that's more complicated to spell than *car*. So, there are ways around the system. This paragraph is in no way an endorsement of that behavior, however.

NEW THC SOURCE

You will often hear the marijuana lobby get shitty about alcohol. After all, why should booze be legal when it has so many worse side effects on the body, and is proven to be so much more addictive to a certain percentage of users? Well, there may have be a meeting of minds between potheads and boozehounds when a new strain of THC hits the market. It's not grown in a field, hothouse, or lab, but made by genetically modified yeast.

Yeast has always been a helper bacteria throughout human history. It makes grains more digestible in bread. It grows naturally on the outside of grapes, helping to break them down and make them more desirable for birds and other animals to eat them and spread their seeds. And it makes alcohol. Sugar and oxygen go in, carbon dioxide and alcohol come out (alcohol is yeast pee).

In a recent podcast, Kevin Smith was astonished to learn that there was such a thing as natural carbonization. But beer is fizzy because of the trapped CO_2 that is a byproduct of the fermentation process. Which is natural. What is *not* natural is using genetic editing to turn yeast into an artificial weed-growing system. Sugar and oxygen go in, carbon dioxide *and THC come out.*

Yes, scientists are fiddling with the natural order of things, which never, *never* goes wrong. So you may ask yourself: Why go to all the trouble? After all, it's called "weed" because it grows, well, like a weed. The reason is purely profit-driven: A small tank of weed yeast can produce the same amount of THC as a one-acre field of marijuana over the course of a growing season. So there's that.

Of course, it doesn't come out of the tank in bud form. They extract the THC from the water and serve it up as a liquid—which shouldn't be a problem, as so many people prefer vaping to smoking, and it is healthier.

THE BENEFITS

Epilepsy is a disorder of the brain that happens on a wide spectrum of severity. My niece had her first seizure in the summer after high school and has had only one other in the six years since then. Still, she can't drive. Dietary restrictions and other ongoing changes to her life are a small price to pay for her safety.

But there are much rarer convulsive seizure disorders on the epilepsy spectrum, and thank heaven they are rare. Not rare enough, though: Sophie, the daughter of our good friends here in Los Angeles, has been afflicted with hundreds of seizures a day since infancy. Some are small, causing spasming. Some are more severe, seizures that last from a few moments to a few minutes. And some seizures, of the grand mal variety, require intense supervision and hours and hours of recovery. The whole thing is terrifying, and the effects are devastating. Her brain has neverhad the chance to develop. Even though she is an adult, she is tiny and emaciated. And theonly thing that helps is pot.

This is not magical thinking. A new study published in the *New England Journal of Medicine*, the first of its kind, focused on cannabidiol, a specific compound in marijuana that significantly reduces the number of convulsive seizures in children with severe and often fatal epilepsy disorder. This is the CBD oil you've heard so much about, which has come into more widespread distribution since two brothers in Colorado developed Charlotte's Web, a low-THC hybrid that is purified into the nonpsychoactive CBD.

The effect on Sophie has been miraculous. Marijuana derivatives have reduced the frequency of her seizures by a factor of a hundred. But you have to try to understand the yoke of distress her family has lived under not only from Sophie's actual condition but also from the wide-ranging and often terrifying treatments that have been suggested to treat her. One doctor wanted to remove half of her brain. And for many years, the best option was to dose her on a cocktail of drugs and chemicals that were not nearly effective enough to justify the negative side effects, including physical addiction—all in a patient who

THE WHY

is not able to discuss how she feels, good or bad.

The studies can't stop here, because CBD is only one compound from marijuana that can be helpful to patients with convulsive seizure disorder. Anecdotal evidence from the web of families living with convulsive-seizure children has shown that if you also introduce THC into the mix, you get even better results. Even looking at THC's proven side effect of anxiety relief shows it can certainly help, since anxiety is yet another side effect of the seizures.

And while we're on that, if you simply look at medical marijuana as a nonaddictive, natural antianxiety medication, in a country where one in six adults take a manufactured drug, it seems like a no-brainer that this kind of treatment will eventually be mainstreamed.

One way to deliver the CBD to the bloodstream is to massage the oil into someone's feet. This is an excellent way to dose a youngster, for example. I recently shared an office with a wonderful mom whose daughter was having night terrors. She would massage a tiny amount of the oil into the soles of her daughter's feet—*as prescribed by a doctor*—and found that this reduced the number of sleep events to about one fifth, before they disappeared altogether and she stopped administering the oil. The only downside is that her daughter loved the actual foot massages, so they will continue.

With so many potential positives, a blanket "Just Say No to Drugs" attitude toward the examination of marijuana is at best ignorant and at worst cruel. We are flying blind here, and both sides of the legalization argument are talking out of their asses. So, do the study and bear me out, Science!

ON THE *EDUMACATION* PODCAST, THE BYE SECTION HIGHLIGHTS NEW BREAKTHROUGHS, TECHNOLOGICAL DEVELOPMENTS, AND OTHER CURRENT EVENTS IN SCIENCE NEWS.

But since this is a book that will gather dust on a shelf for years to come, it made sense to us to change this section into a look at what we might expect from science in the future—or at the very least, to show how predictions and solutions are shaping up now.

Don't expect Nostradamus-style prognostications here. While I do spend a lot of time staring into a pool, it's usually just an attempt to find all those lost earring backs our guests keep losing when they come over for a swim. Also, unlike Nostradamus, I strive to minimize the mistakes: Thanks to a crack squad of fact-checkers, you won't see any mention of Adolph Hister in these pages, I can guarantee that (except for that one just now).

I have to bring up James Burke, author and host of the 1970s BBC series *Connections*, which took all the technological building blocks from over the centuries and showed how—when combined in clever ways—they created the great technologies that changed society. I love this man's style: Rather than just describe the kind of damage a broadsword did to someone on the battlefield, Burke used one to hack apart a side of beef—right on screen!

I bring it up again because *Connections*, like most of Mr. Burke's work, was optimistic. Sure, there was some talk of nuclear holocausts and so forth, but this was years before Reagan and Gorbachev and if this idea wasn't in the show, it would've been conspicuous in its absence. The show really celebrated ingenuity and awesomeness. And Burke kept it going with more projects like *The Day the Universe Changed* and the ambitious Knowledge Web, innovative in their own right. He is a celebrated lecturer and futurist, always looking ahead in a positive way.

But lately he has had a darker view of what might come to pass for humankind. In a recent interview by Matt Novak on his podcast *You Are Not So Smart*, Mr. Burke predicted that mankind is very likely to destroy itself—and not through the classic struggle over land or religion or any other conflicts that have plagued us from time immemorial.

In short (and really not doing Mr. Burke's brilliance any justice, I might add), those conflicts arose due to scarcity and how we deal with it. We have religion to try to hedge against scarcity, in the sense that one worshipped Pharaoh under the condition that he brought the crops back each year. We fight over land so we can have the other guy's stuff. But with the developing ability to manipulate matter at a subatomic level, we will be able to create anything, and scarcity will disappear. And so will the social contract, since all the rules that govern how we deal with each other on every level will go out the window.

Any nut, Burke says, will be able to shape his entire world in whatever image he likes. Or create a superweapon and use it. Or manipulate life into whatever horrible deranged form his mind can imagine. It will be like CGI for real—and like in the movies, we will have just as many horrors as we do superheroes.

But let's not worry about that for now! Let's worry about all that other crap that the nightly news and scaremongers are saying will ruin things for everyone! There are a lot of dumb ideas out there to save the world, so stop worrying and enjoy!

And just for you, I managed to get through a discussion of the future without mentioning global warming once! (Except for that one just now. Dammit!)

Chapter 25

DUMB IDEAS
TO SAVE THE WORLD

If variety is the spice of life, then the wonderful variety of existential crises facing the human race makes life *extra* spicy. But since this book is nominally about science, we will set aside all the fun fears from the nightly news, like rogue nuclear nations and the preponderance of side effects from the drugs advertised during the commercial breaks.

No, let's talk big-picture here: global warming, big-ass asteroids (or big-assteroids, as nobody wants to call them), mega solar flares, the sudden reversal of Earth's magnetic field, rogue black holes (as opposed to the law-abiding ones?), global epidemics, and so on. There just aren't enough three a.m.'s to go around to worry about them all.

By a much higher standard than that set by Alanis Morissette, it is ironic that predicting the end of the world used to be a mostly religious practice, but now science has the cultists beat by a mile. I mean, you have to go to school for years to come up with flood-basalt volcanism, which describes the 1783 eruption of Iceland's Laki volcano, which killed nine thousand people off the bat, caused a famine that killed a quarter of the remaining population, and blacked out the sun, lowering the temperature that winter by nine degrees Fahrenheit *all over the world*. It's not enough just to say, "Big mountain go boom" anymore; science understands the world's complexities enough to put a complicated name to it and then make a prediction.

And perhaps even *more* ironic than that, scientists are thinking of ways to stop these world enders in ways that make less sense than tossing a couple of nice kids into the molten lava. I get it: To think ahead on a global scale, you gotta think *big*. And by *big*, I mean "dumb." Because some of the solutions rely on technology we don't even have. Or materials that don't currently exist. Or expertise no one possesses. Or even bending the laws of physics. I mean, I get that mathematicians want to save the world, but when you look at the recent claim by a theoretical mathematician that he has solved how to travel back in time, and he has named the argument TARDIS, you might start to think, as I do, that a large, theoretical grain of salt should be taken here.

My criteria for this chapter: The threat has to be imminent, the solution doable, instead of just sounding like fan fiction based on a math book instead of *Twilight*.

So, let's not focus on the bad things that might befall our species. Let's look at the bad ideas that legitimate scientists have come up with to keep the human race limping along until it either destroys itself or the sun dies out.

ASTEROIDS

It has been estimated that humans have spent more money making movies about trying to stop interplanetary collisions than they have spent actually trying to stop such a cataclysmic event. So, if you don't remember the movie *Deep Impact*, then obviously it did not have a deep impact on you. And there is a lot to be learned from *Armageddon*: For instance, NASA would probably have a lot more money lying around if they started doing product placement. And they certainly would be a lot more interesting and fun it they were as attractive as, and drank as much as, the cast.

However, what we should *not* use as a takeaway from *Armageddon* is that Michael Bay is a good director who looks after his stuntmen, and that if a big-ass-teroid (I'm sticking with that) comes along, we should blow it up. With a bunch of roughnecks from an oil rig. In space.

Blowing it up would just cause a shitstorm of debris to fall to Earth, and while some of the mass of the original rock would be deflected, the operation of blowing it up would have to take place in such close proximity to Earth that it might not only cause more widespread damage, but those chunks would also be radioactive for the next 10,000 years or more—so, yeah, your descendants are alive, but they get kidnapped to serve in the Pre-Crime psy-ops pool.

What you gotta do is land on that rock and *nudge* it out of the way while it's still far enough away to get out of Earth's path. Sound crazy? Well, the European Space Agency landed the Rosetta probe on a comet not only to teach the comet Spanish so it could spend the summer in Europe with its girlfriend but also to study the composition of the comet—and, frankly, to see if we could do it. And we did! And by "we," I mean the European Space Agency.

So, is it a good idea? Not yet. The Rosetta was able to perform only a few functions and transmit a little bit of data, since the active ingredient landed in a shadowy recess in the comet's surface and ran out of battery power. With only enough power to watch one episode of *Mike Tyson's Mysteries*, there's no way that probe would be able to push that comet. And while a comet can be much bigger than a deadly asteroid (as big as 6 miles across), it lacks the mass, as scientists say that most comets are simply "dirty snowballs," which sounds like the name of a pornographic Icelandic ice-sculpting competition.

So, while everyone seems to be in agreement with the European Space Agency that this is the right approach, the only way we can make this work would be if the asteroid first flew by Earth and then politely circled back to destroy us, so we'd have time to figure out how to intercept, land on, and move it.

GLOBAL WARMING

There is a lot of ignorance surrounding the global-warming discussion. Much of it is the worst kind—people who refuse to look at the evidence gathered, collated, and presented by the very people we pay to understand it. Well, used to pay. These days, if you're a scientist working for the government, I think it would be overly optimistic to iron five shirts on Sunday.

One thing that really bugs me about climate deniers is that they think this is the first time in the history of Earth that man has created some kind of catastrophic climate change. Not true! After the conquistadors wiped out much of the population of Central and South America, there was a period of global *cooling* as millions of acres of abandoned farmland were taken over by forests, thus taking carbon *out* of the atmosphere and allowing more heat to escape the atmosphere and causing the Little Ice Age (which, thank God, really happened and wasn't another annoying CGI-animated movie).

And don't forget the Kuwaiti oil fires, set alight by Saddam's fleeing troops during the first Gulf War in 1991. *That* was a very efficient way to get CO_2 into the atmosphere, as well as every kind of poisonous fug you can imagine. While the environmental impact was serious, if it hadn't been for Red Adair and the Texas Troubleshooters knocking the fires down within about two years after they were ignited, the impact on the environment would have been devastating. Carl Sagan estimated that after only one more year, we could have seen the same results as the Tambora volcano explosion of 1815, where so much sunlight was blocked that it became known as the Year Without a Summer. Winter is coming, indeed!

So, if we can stop blindly denying the problem and focus on solutions, what innovative (read: *idiotic*) ideas are out there to reverse the effects of climate change? The one that's getting the most play right now comes from actorly named MIT scientist David Keith. He proposes fitting Gulfstream business jets with sprayers that emit a fine mist of sulfuric acid, then flying them around equatorial regions at 13 miles up. The resulting sulfate aerosols would infuse the stratosphere around the world and reflect the heat. Just eleven jets could reduce global

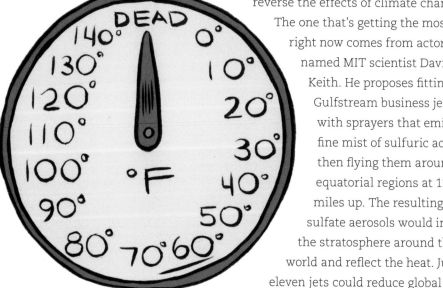

warming by more than one degree per year. The only problem? It might be irreversible! Whoops!

But here's the worst one: A number of scientists have proposed putting either a giant mirror—or lots of little mirrors—into space to deflect about 2 percent of the sun's rays from hitting the planet. Unfortunately, that would take about a hundred years to implement, either in manufacturing the big mirror in space or deploying three *billion* mirrors, and would require about five million space ships on constant round trips. And at least one hundred ships would have to be replaced or repaired *every day*. By the way, using conventional launch techniques, the fuel expenditure would be twenty times the global rate and would in itself cause the worst environmental disaster in history.

Two articles on this method ended with "This approach needs more study." Ya think?

MELTING ICE CAPS

Yet another argument put forth by global-warming deniers is that a few degrees in average temperature isn't going to make a big difference. But it makes a huge difference!

Now, I know I'm talking about stuff that usually ruins the *Planet Earth* episode for everybody, and hopefully you know all of this. But, c'mon, if you don't, please, please be alarmed, and know that anyone with any interest in science—and most politicians, for that matter—have either sold their beach houses or bought a shit-ton of sandbags.

Because global warming is melting the ice caps. Which is bad for so many reasons, but I'll name only a few here: The ocean levels will rise, causing flooding and displacing (i.e., making homeless) at least a half a *billion* people. Which will put a lot more people out, because it's also going to flood croplands with seawater, making that land infertile. Northern areas with extreme tides will see even *more* extreme tides, changing ocean-flow patterns and potentially killing deep-sea animals—those that are not dead, that is, from the dropping oxygen rate, thanks to the algae blooms associated with coastal flooding. And with an extra few feet of ocean overhead, many coral species will die, devastating the marine population they support.

Even if you live far above sea level in the Rockies, for instance, the melting ice caps can get you: They are discovering bacteria and viruses that have been frozen in the ice for millions of years just waking up and saying hi—and not in a nice way. More in a John Carpenter's *The Thing* way. And, no, they won't be easy to kill just because they're old. A recent study showed that the superbugs that are making hospital stays so dangerous predate the dinosaurs. Hand sanitizers and antimicrobial soaps are just killing off their competition, allowing them to thrive.

So, short of stopping global warming, which brilliant solutions are being proposed to stop the resultant melting ice caps? Blankets. Yes, *blankets!* It's a simple idea: Just like you pull that comforter over you to keep the cold out, a thermal layer called an ice protector—which means "blanket" in Switzerland—lies atop the ice, protecting it and keeping it cold. The reason I mention Switzerland is that commercial ski resorts in the Swiss Alps have been using these for more than a decade to protect their snowpack—and thus investment—from melting in the off-season. And heaven knows, the lord Alps those who Alp themselves.

When you look at the scale of this solution, you realize it's a lot closer to the three-billion-mirror way of thinking. Even though a large thermal tarp was able to allow a small glacial formation to retain 80 percent more ice compared to neighboring (and much larger) glaciers, that's pretty small potatoes when it comes to gift-wrapping the top and bottom of our planet. To give you an idea of the scale, you know all the plastic waste in the ocean that everyone is so upset about? If you spread it in a 1-inch layer over the polar ice caps, it would cover only about one-sixteenth of the area needing protection. And the job would make *Ice Road Truckers* look like *Cupcake Wars*.

ROBOT APOCALYPSE

Of all the apocalypses, this is my favorite. So much better than the zombie apocalypse, which has already taken over several time slots on AMC. Instead of having some drooling, extremely gross-looking shuffling moron take a bite out of me and turn me into a drooling moron (who will probably be grosser-looking, since I'm not starting out so strong in the looks department), I imagine C-3PO politely harvesting my organs for some terribly important reason that just happens to escape him at the moment. But really, it's a Sophie's choice, and I would actually prefer not to be apocalypsed on in the first place.

Of all the questions we get for the *Edumacation* podcast, the most popular one is, "When will computers become smarter than us, thus ending the world?" The jury is out on that one, both in terms of how you measure intelligence or whether AI is even possible at all. And it can be argued (by way of the current political climate) that computers have already surpassed at least some members of the voting public in terms of smarts. It is too easy to think that politics is rapidly becoming like *War Games* with Matthew Broderick: The only way to win is to not play.

While the question of whether machines will eventually dominate man seems paranoid, consider this: Privacy is one of the top principles that conservatives and liberals agree should be kept as a right, even if it requires more legislation. And yet we share *everything* on social media, and how that information is exposed and how we are exposed to other friends' photos, stories, and news articles is optimized by machines. Add to that the fact that we are rapidly adopting technology that is constantly listening to us—if you own an iPhone, just say, "Hey, Siri!"—and you might not be *that* paranoid that there could be more to Alexa than playing music and ordering coffee beans. The Internet of Things— smart-connected appliances and so forth—are also listening (and vulnerable to cyberattack). And technically, since my smart fridge can analyze its contents and restock itself, it is a machine that can perform tasks unsupervised; it's a robot. Does this mean my blender is going to rise up and destroy me? Well, it's never really had a shot until now, so why risk it?

So, fast-forward: There's an army of androgynous robots voiced by Alan Tudyk, and the only thing between them and total global domination is Will Smith, a gritty cop who does his thinking with his fists and loves his son—or some random kid he knows; it's not clear. What do you do?

Here's my solution: You can disconnect. Take all your electronics and throw them into the smoking remains of the Genius Bar, which will be destroyed not by robots but by customers who never got their stuff fixed and resented having to call those guys "geniuses." Go live in a Faraday cage—the mesh in a cage will stop many electronic transmissions from penetrating. If you are a Sprint customer, you will be familiar with a Faraday cage, since it seems like there's one covering most of Los Angeles.

Next, I'm assuming patient zero in this scenario would be some kind of combat machine built by DARPA, so it's pretty likely you'll lose in a one-on-one situation. I have zero patience for patient zero, but there are ways to get under its skin, so to speak: Get your buddies together and go to your local hospital and steal that MRI machine. Its name stands for "magnetic resonance imager," and that giant tube you slide into before the "bam-bam-bam" starts is a giant electromagnet, with *some* shielding, that you and your buddies can remove.

Put that puppy on your diesel truck (I'll tell you why in a second) with a large generator. I recommend the WhisperWatt 9000, because it's nice and quiet and you'll be able to hear their robot screams. Now, you and your pals start driving around town, firing up the MRI every dozen feet or so. You will have knocked out all electronic devices with that EM field in no time. (Why a diesel? Because it has glow plugs that stay hot, instead of spark plugs, which need an electronic signal, so you can drive that thing from here to your nuclear-winter wonderland if you want.)

And if you fail? Well, you're already living in a cage by choice—how much worse can it get?

Chapter 26

PRODUCTS FOR A STUPID FUTURE

I was lucky enough to work on a show for Chris Hardwick at NBC that was all about future technologies and how awesome everything is about to get. When pitching shows, often, the producers or executives will produce what is called a sizzle reel—images and music, sometimes narration or subtitles, that convey a feeling, or give a slice, of what you can expect from the show. I might have seen that sizzle reel ten times, and I cried every time. Because the future looked awesome!

There was that kid who got new arms sewn on. Now he plays baseball! There was that formerly blind baby who saw his parents for the first time. (It plays on my emotions like a concentrated shot of the Hallmark Channel!) By the looks of things, the world—the whole world—is going to be a much, much better place.

But after working for a few days, I realized that there were a lot of very stupid ideas out there that people were pursuing just because they could: The Star Wars enthusiasts who converted two water jet packs to be like speeder bikes from *Episode VI.* Cool, but, you know, putting the whole future of the world on hold. Or the doctors working with miraculous CRISPR process (see Chapter 27, "A Better Tomorrow") who really are using it to make people forever young, instead of just giving them plastic surgery to *look* young. Or the amazing 3-D food printer, which can store basic ingredients for years and make a wide range of foods—and is printing out pizzas all the time. (Because if you want to put someone out of a job, make it the pizza guy!) Or the process that can actually take CO_2 out of the air and turn it into ethanol—so we can keep driving internal combustion cars and put the CO_2 right back where we found it.

So, yeah, there's a line. On one side, you have the men and women who are working to make the world a better place, and on the other side, you have the men and women who are working to make their mansion a better place. Nothing wrong with that—in fact, there's something great about those who are developing technology to get rich: You can make fun of them. I mean, how the hell am I supposed to take the guys making artificial blood down a peg? It's impossible! They could be huge fans of the latest *Pirates of the Caribbean* sequel, and I'd still have to give them a pass.

Another difference is that on the altruistic side of the line, you have people working to do or make something they know the world needs. Clean water. Less waste. Protection from disease. Better communication. And on the other side of the line—the preposterous venture capital side—you have people who are doing or making things they think people will buy. One side of the line is always right. The other side is usually wrong.

I wish I had room for all the ideas that will end up on the dust heap of history. Like the chalk recycler, which trolls up and down the base of a teacher's blackboard, vacuums up all the chalk dust, and forms it into a new piece of chalk. What a great way to drain eight C batteries in the process, because it's much cheaper and more convenient than having to buy chalk! Or electromagnetic dumbbells that automatically adjust their weight so you don't have to bend down and get a different dumbbell, *thus getting all that unnecessary exercise.* Or the maglev-cloud couch, which floats above the floor in a magnetic field like a cloud but feels exactly the same as if it were on the floor, only now the EM field has fried your iPhone.

I must mention at this point that, again as part of *The Awesome Show*, I attended a two-day conference by a well-known tech incubator (or "accelerator," as they are now buzzworded) in Silicon Valley. In addition to having the least television-ready group of presenters I had ever seen (unless *American Horror Story* is planning a season called *Tech*), they gave us a very rosy—if skewed toward their own companies—view of how tech is going to make everything better and drones will deliver our pizzas and soon healthcare will be like having a check-engine light for your body. All of these ideas were very facile, and meant to persuade us just what was cool. The Internet of Things being one example. So let me be clear: While the *future* of this stuff might be great, the way it exists *right now* is what I'm talking about here. I hope to live in a world where all this shit comes to pass and makes our lives better, but for now, let me just react.

Let's take a look at some more of the wrongheaded products and misguided ideas that someone, somewhere is convinced will make the world beat a path to their door.

BRAIN-TO-BRAIN COMMUNICATION

There have been a number of significant advancements in the field of brain-to-electronic communication. Most of the breakthroughs have come in the form of assistance to people with spinal injuries. Electrodes in the brain receive impulses from the motor center that used to control a limb, for example, then that electronic signal is sent and interpreted by a hardware interface that controls either a robotic limb or some kind of exoskeleton or mobility aid, literally allowing the user to take back their independence with their mind. I applaud these efforts; it really is a beautiful thing to see someone who is able to make their way on their own again.

Researchers even stepped up their game, letting people communicate through brain impulses alone. Last year a woman, nearly immobilized by multiple sclerosis and unable to speak, was given a similar rig that takes impulses from the part of her brain's motor center that controls her fingers.

After working with the interface for a few months, she is now able to imagine typing on a keyboard, and the interface displays the words she is thinking, letting her talk to her family, write, and even perform word commands to control her mobility devices. Awesome!

But someone had to take it too far. Researchers at the University of Washington have used electronic brain interfaces to allow one scientist to control the hand movements of another scientist using only his mind. And the interface. And of course, DARPA, the oddball US Defense Department research arm, wants to develop an interface that will allow soldiers to communicate telepathically on the battlefield. Because telepathic soldiers do not conjure a totally nightmarish vision of the future.

Helping injured people is great. Installing a USB drive in a person's brain so they can silently kill, not so much.

SELF-DRIVING CARS

Pop quiz, hotshot: How many of the major companies developing self-driving cars have had incidents where the car has hit a pedestrian? Answer: all of them.

Silicon Valley meets Detroit in what could be the biggest boondoggle the auto industry has ever seen. Not that self-driving cars won't eventually come to exist. It's that they are racing to get it done, wasting time to get to an artificial finish line that many consumers simply don't want. However, it's a pretty sexy item for the company manifesto, and when you look at services like Uber, it makes perfect sense: They have been dehumanizing their drivers since the jump, blaming things like tens of millions of dollars of underpayment to New York drivers on a software glitch. But even the drivers find the self-driving-car idea exciting.

One Uber driver we had in Washington, DC, recently told us that he has already started making payments on a new Tesla SUV that not only doesn't exist yet but also will drive itself on a system that hasn't even been designed yet. But he's pumped, because once his $175,000 limo-SUV rolls off the line, he will be raking in some sweet, sweet coin while his car—not him—does the earning.

Here is the scenario they have pitched him: Say he wants to go out to dinner with his wife. His Tesla will drive them to dinner, but they don't need to pay for a valet, because the car just drives off and finds a parking space for free. While they're not using it, the car is on call, and if someone wants a ride, it goes to pick them up, first stopping to pick up a "supervising driver," who sits behind the wheel in case human control is needed. The car picks up the passengers, drives them to their destination, then drops off the driver, presumably at the homeless shelter or outpatient clinic from where it picked him up. It then parks close to the restaurant and waits for the Bat-Signal to arrive, then picks up the owner and his wife.

That sounds astonishing, amazing—and totally stupid. The owner gets a piece of the fare, the service gets a piece of the fare, the driver gets a piece of the fare. Like always. Only, the driver doesn't do anything and the car is out driving twice as far as it needs to go, and when the owner is done with dinner, he and his wife are three times as likely to encounter a stranger's fart than if they had just left the car with the valet.

You know what we need? Noncrashing cars. Then any idiot can drive one. But a self-driving car? It would be easier for this guy to just buy a car, hire a driver, and have him drop them off at dinner.

THE INTERNET OF THINGS

Somehow it happens that our future is going to be full of so-called smart devices in the home. I get a smart TV: It is already a sophisticated piece of electronics that interfaces with various methods to deliver entertainment content to the home. Maybe someday the assholes who made the brain-to-brain device will let us control our TVs with our minds, thus eliminating the incredible inconvenience of moving your arms and fingers to operate the remote. But a smart toaster oven? A smart fridge?

As I've said before, even people who are up in arms about privacy issues voluntarily install devices that, in effect, spy on them. And now you can get a refrigerator that tracks the food you keep in it. There are some practical applications for this, like reading the sell-by dates to alert you when it's time to toss that yogurt. Or reading the levels of various products like milk and the fruit in the crisper, generating a shopping list, and texting it to you. But the goal is to have it be much smarter than that: It will learn your habits, tracking and scheduling how much you eat and drink, then coordinating with other smart devices like your fitness tracker to determine whether you are expending enough calories to justify all that flan.

There is a wonderful web-based service called IFTTT, which stands for "if this, then that." It allows you to connect your smartphone, home controllers, vehicle chargers—anything you've got that works in service of your needs, has an Internet connection, and keeps a record. Why connect? Because then things get really interesting. You can make it consult the National Weather Service and have it always turn on your lights exactly at sunset. You can have it monitor your phone's GPS and have it start the air conditioner when you're a certain distance from your home. You can have it monitor who's calling you so that if it's someone you don't like, then all the lights in your living room turn red. There are a million ways to combine your security cameras with your heating system, and now your toaster.

But, as with nearly every online service, there is usually a bit of fine print that says your user data may be anonymously monitored to improve customers' experiences. I know it sounds paranoid, but if one of the thousands of data-mining companies can get your income based on your ZIP code and your birthday, how much more info do you think they're going to get from your GPS coordinates, and twenty devices with live mics, and cameras distributed throughout your home?

In light of recent Russian hacking scandals that disrupted elections and brought down banking servers, do you really want to pay a ransomware fee in your own kitchen just so you can eat some leftovers?

Of course, the idea of the Internet of Things is not limited to consumer items. It also applies to any smart devices that work together without human interference. I recently attended a demonstration at Microsoft of a system that uses drones, ground sensors, and a constantly evolving AI to optimize the yield of small, independent farms—low-cost tech that enables these independent growers, most of whom are born into the work, to stay competitive with large farming conglomerates. And if we ever get flying cars, there is even a system being developed in the San Francisco Bay Area to have drones flying in formation, creating drone roads in the sky that dynamically form along the sides of your vehicle's three-dimensional path, then break up and form new roads for you or other drivers as you pass. They create a visual reference for the driver—or self-driving flying car.

DARPA (DEFENSE ADVANCED RESEARCH PROJECTS AGENCY)

I brought this up early with the silent-soldier scenario, but let's drill down a little deeper and explore where this nightmarish skunkworks might lead us.

To be clear, I support our troops, and I applaud any kind of creative thinking that will keep them safe and out of harm's way. And DARPA does a lot of that well. They have a new carbon-fiber material that is lighter, twice as strong, and much stiffer than the old carbon fiber. They use it to make things like Geckskin, a material that allows soldiers to cling to and climb up the side of a building like geckos. They are developing a missile that will launch a satellite into space from the underwing of an F-16, drastically reducing costs and allowing extremely fast deployment. GPS, the M-16 rifle—hell, back in the 1960s, when it was just called ARPA, it came up with the Internet (or ARPAnet, as it was known and never called).

But, hoo boy, have they got some harebrained and dangerous ideas. More to the point, they may not be harebrained ideas: The harebrained part is that they are actually pursuing them. I'd like to suckle at the government teat of military research spending as much as the next guy, but my conscience took enough of a beating as a monologue writer; I'm not sure it could make it through one of these projects.

They want to modify insects by turning them into cyborgs, adding electronic controls and enhancements to make them remote surveillance cameras and even allow them to coordinate to become killing machines. What is wrong with these guys? Don't they watch *Black Mirror*? A swarm of cyborg killer bees? What could possibly go *right*?

They are developing what they are calling "programmable shape-shifting matter," which, yes, could be something as innocent as glasses that auto-matically adjust to your prescription or antennas that change shape to optimize signals on different frequencies. Or it could be something more sinister, like a tie clip that kills you, or a desk lamp that kills you, or a class ring that morphs into a drone that leads a swarm of cyborg killer bees (to kill you). Come on, guys, *Terminator 2* wasn't scary enough for you?

And speaking of *The Terminator*, here's the scariest idea of all: Yes, we have drones that allow pilots on the ground to remotely target and destroy almost anything with incredible precision and very little risk, and history will judge whether this made Obama a good president or a bad one. But at least someone is controlling them. The newest nightmare from DARPA is an autonomous robotic soldier that will make discretionary kill decisions in the field, without the need for human interface. The fire-and-forget approach—except this thing doesn't just explode at the end of its journey, it goes tear-assing through town all day deciding who lives and who dies.

DARPA! Haven't you ever seen *Robocop?* Or *Weapon?* Or *Westworld? Killer Klowns from Outer Space? Metropolis? Logan's Run? Alien?*

DARPA should take part of its massive budget and get a goddamned subscription to Netflix!

Chapter 27

A BETTER TOMORROW

A theme that is an absolute necessity in capitalism is that life is going to get better because our stuff is going to get better. Better living through chemistry. Higher-resolution television monitors. Faster Internet. But if you know anything about me, you know that I'm setting up a cautionary tale. And if you don't, read that last sentence again, because I'm setting up a cautionary tale.

That technology improves quality of life is certainly debatable. Consider the mobile phone. Back when they were 2-pound handsets attached to portable car batteries, and using them could shut down the delicate monitoring equipment in hospitals, it really was the best symbol of what would be great about the future. The Starfleet uniform insignias–slash–two-way radios on *Star Trek: The Next Generation* were a futuristic vision of how a personal communicator could provide new levels of convenience and service. But, alas, this was pure fantasy, because Jean-Luc Picard's man-brooch didn't turn into an electronic leash that allowed anybody to get in touch with him at any time. ("I'm battling a Moltaran slime creature. Can I call you back? Is this important?")

But where the mobile phone was something of an inconvenience, the smartphone was a behavioral and societal game changer that has definitely made the world worse. These phones enslave us. *Time* magazine reported that the average person checks their smartphone forty-six times a day. After they post on social media, that number goes way up as people ache for that dopamine rush of getting likes. Stand in line at the bank, and you'll see that everyone is looming over their smartphone screens, actively engaged in something that's either happening remotely or in a game or activity on their pocket computers. The flesh-and-blood part of the person has become more like an avatar than their online persona. *HuffPost* keeps trying to attach derisive names to these new social actions, like *phubbing* for "phone snubbing," but millennials are all *fuck HuffPost* (which gives me hope for the millennials).

As I mentioned above, James Burke was the creator and host of *Connections*, a 1978 show—and, later, book—about how technology draws on past achievements to create new innovations. It's what got me interested in, well, *everything*. He went on to write many fascinating books about science and technology and is a sought-after lecturer and futurist. His version of the future has always been tied to how technology changes our lives, and it is usually for the better. But as he lives through his twilight years, things are taking a dark turn.

Recently, on Matt Novak's *You Are Not So Smart* podcast, Novak caught up with Burke after a lecture and asked him about the future. And it didn't sound good, because up until the 1980s, the linear progress of technology had an overall positive trajectory. But now the most significant tech is disruptive. No one could have predicted the profoundly negative effect that the smartphone would have on society—which can shake up a futurist—yet that's nothing compared to disruptive fabrication technology.

Take your 3-D printer. I have one, a cheapie that was a pain in the ass to tune but at least let me crank out the mini pickle helmets the Bratzis wore in *Yoga Hosers*. It's basically a very small hot-glue gun that's controlled by a computer. The computer slices up a 3-D model like a CAT scan, and then a little hot tip melts plastic and deposits it in one long strand, like a 3-D rag throw rug. All the technology has been around for decades, but even so, print out a poop-emoji key ring, and you are guaranteed to get a "Holy fuck!" out of most people.

Wanna get *really* astonished? Several companies, including AT&T, are developing fabricators that manipulate material on an *atomic level*. Meaning you feed in whatever material you want, and it can *change the elements themselves*. You can make anything. Maybe even living tissue. And the first thing you'll do is make another fabricator. Within months, everyone on Earth will have one. And then the world will come to an end, because our society is based on scarcity.

All politics, the social contract, the law—all of it exists so that people don't go out and take other people's shit that they ain't got. Scarcity makes us negotiate, be reasonable, find common ground. Now, imagine that anyone in the world can have anything they want, from wealth to their own version of people to their own atomic weapons. Diplomacy goes out the window, and mankind disappears in a puff of progress.

But this is a fun book, remember? So, let's look at products and innovations that are meant to make our future better and brighter, starting with the everyday and working toward the truly miraculous.

SNORING

As long as there has been man, there has been snoring. I only assume this—I have no anecdotal evidence beyond having a college roommate who spent seven hours a night sounding like someone trying to start a chain saw with no gas in it. But despite the long-standing nocturnal vibrations of relaxed respiratory structures, there has never been a satisfactory solution to the problem. And believe it or not, solving it could improve your health and lengthen your life. By a lot.

First up: the CPAP. Stands for "continuous positive airway pressure." It's used in bedrooms around the world to force air past that throat bassoon and keep the person breathing. It's also used in movies and TV to show that a character is buffoonish or pathetic in some way. The CPAP is very rarely prescribed for snoring—its usual application is for sleep apnea, which is like super-snoring that can cause heart damage. But just as the standards for cholesterol levels have lowered due to pressure from drug companies, the standards for what is classified as sleep apnea are coming down, so you're going to see more and more people slipping a tube into their mouths before bedtime. And then they'll put on their CPAPs.

Next up are the dental appliances, which come in a few forms. The one I use looks like a modified, clear pacifier, but instead of a bulb in your mouth, it is open ended and large enough to hang onto your tongue. You squeeze it slightly, your tongue gets sucked in, and you close your mouth around the wider end, making your stick your tongue out all night, but also tightening the soft tissues that vibrate. (The morning breath is astonishing.) Other appliances hold the tongue to the floor of the mouth or otherwise keep the tongue from falling back into the throat. The US Army uses this kind of solution for barracks living and field maneuvers.

The standby surgical option is called uvulopalatopharyngoplasty. (That's a word I have been waiting all my life to somehow manage to get into a book. It's why I stopped writing mysteries—very hard to work *uvulopalatopharyngoplasty* into a crime scene.) Basically, the doctor removes tissue from the back of the throat, but it's risky, because you can get scarring that actually reduces your airway size and can be dangerous.

But if you must have uvulopalatopharyngoplasty, there's a new technique that is super-cool and futuristic and much safer than the old uvulopalatopharyngoplasty. It's called radiofrequency ablation, and it uses radio signals to intentionally create scars under the skin in the throat, which tightens up that throat skin better than Archie Bell and the Drells. However, it's permanent, and only reduces snoring, instead of completely eliminating it.

My money's on the Pillar Procedure. Not only does it sound like the title of a Len Deighton novel, it is a minimally invasive, reversible surgical procedure. Basically, they implant suture material in the back of the throat, which stiffens it up more or less, depending on how much material they insert. Thus, it is customizable, and can be fine-tuned to specific areas of the soft palate to maximize effectiveness. You take away all the health risks in less than thirty minutes, and the process doesn't have any side effects—except for earning you the undying love of your spouse/partner/goldfish/neighbor/guys in the surveillance van.

HYPERLOOP

You can't write about the future without mentioning Elon Musk, because in the future, he invented a time machine and is standing behind me right now with a plasma disintegrator to my head.

For more than fifty years now, graphic artists have been making beer money with concept drawings of maglev, or magnetic levitation, trains. And, yeah, they go fast and are efficient and all that, but they aren't enough to be game changers, because they are just glorified bullet trains, are subject to bird strikes and environmental hazards, and, like most transportation technologies, spend most of their energies breaking the wind. (Insert crazy-uncle joke here.) More to the point, even though they are better for the environment, they just don't go fast enough to rival air travel—the Shanghai maglev train tops out at 267 mph (430 km/h), well short of commercial airliners' typical cruising speed of 586 mph (945 km/h). Most of the actual time savings comes from proposed maglev tracks directly getting you from city center to city center, requiring tunnels and all sorts of headaches.

"But if you're going to have to tunnel into the city centers," says Musk, his spandex-clad finger playfully squeezing at the plasma disintegrator trigger, "why not enclose the entire process in a tube?" And so, the Hyperloop.

It's a maglev train inside a tunnel from which the air has been evacuated. This will allow the train to travel at speeds of over 800 miles per hour and be completely protected from environmental hazards. This will get passengers from downtown San Francisco to downtown Los Angeles in thirty minutes, but will do nothing about their disappointment about leaving downtown San Francisco and ending up in downtown Los Angeles.

Proposed to travel just over Mach 1, it doesn't produce a sonic boom, because there's no air to carry the pressure wave. But at that speed, it's not like you can just turn a lot of corners, so either the path of the train is going to have to be underground, or he's going to piss off a lot of Central Valley farmers by getting Sacramento to exercise eminent domain. Plus, the cost is mind bending.

Wait! I'm sorry, Mr. Musk! What I meant to say is that it's a great idea and it's going to happen! It's definitely going to happen!

FLYING CAR

According to a fascinating web series produced by the *Smithsonian*, the *Jetsons* cartoon was among the most imaginative, predictive, and influential visions of the future of technology ever presented in popular culture—the active ingredient in that sentence being "popular culture." We never knew we wanted all those things, but now, in the average home, think about how many of those inventions we have: push-button coffee makers, screens that let you see who you're talking to on the phone, CCTV of who's at the front door, dog treadmills, a machine that cuts your hair with a vacuum cleaner, jet packs, a tiny boss who screams at you all the time.

So where's the flying car?

Kids, you're in luck. As of this writing, the *New York Times* has reported that Silicon Valley has taken over the problem and is rapidly solving it. Google cofounder Larry Page is funding the Kitty Hawk Flyer, a personal transport vehicle that you can fly anywhere and land in a regular-sized parking space. After kicking up dust and dirt all over the other cars. If the flying car does one thing, it will be to put the people who use leaf blowers out of business.

There have been, by definition, many flying cars throughout the years, most of which are small airplanes with retracting or folding wings that allow them to drive on the freeway without taking the lights off police cars. What's bad about this idea is that you get a vehicle that is both a slow and unsafe airplane and a slow and unsafe car.

The new generation are VTOLs. That acronym stands for "vertical takeoff and landing," and most of them rely on the same basic technology as a hobby drone. A number of propellers or ducted fans lift you up, level you out, propel you forward, and set you down based on some simple controls. You can find videos, though you will notice they're always filmed over open water, because if that thing fails to do any of the steps listed above, it's nicer to hit the drink than drop onto a tai chi class in Palo Alto—though not by much, because it would be hilarious to see how slowly that tai chi class runs from a plunging flying car.

I want to add a caveat here, because I want a flying car, but the whole idea will succeed only if a) there's a market for it, b) it's safe, and c) you can get insurance for it. Having it exist doesn't necessarily mean you will ever have one, and I'm not talking about the cost, which will be high. While the FAA makes it remarkably easy for anyone to take experimental vehicles into the sky, it takes only one massacred tai chi class to start putting restrictive regulations in place.

And sometimes ideas just don't catch on: In 1992, a man in Manhattan Beach, California, invented a small personal helicopter with a blade diameter of 8 feet that could safely take off and land in an area half the size of a tennis court. He went out of business. It turns out, in a city with the worst traffic in the United States, people would rather sit in traffic than drop out of the sky and kill some slow-moving people just trying to get a little exercise.

Caveats aside, this is really happening. Flying cars, like talking refrigerators, self-driving cars, and unattractive wearable technology, will force their way into our lives if Silicon Valley has anything to say about it. And especially if we lay out a few bucks on Kickstarter.

THE ONE-SHOT CANCER TREATMENT

If you've been paying attention to humanity recently, you may have heard of Emily Whitehead, a tween girl who is in a very small club of cancer survivors. She had acute lymphoblastic leukemia, which is a common childhood cancer, but was among the unlucky few who did not respond to traditional treatments, and found herself in stage 4. The "small club" I'm referring to is the relatively few patients who have had stage 4 cancer and who not only went into remission, but who were *completely cured and are now cancer-free*!

The treatment, called CAR-T therapy, is a last-ditch, Hail Mary pass at getting rid of the cancer. It uses several methods found in other cancer treatments, like bone marrow removal and re-transplant, as seen in multiple myeloma treatments that my friend went through for a year in Little Rock, Arkansas. But unlike that and other efforts, this is a one-shot deal.

The reason the cancer grows in these patients is that it disguises itself from detection by their immune systems. But there's a way around that. In short, the doctors extract T-cells—immune cells and bone marrow—from the patient. Then they re-engineer them using genetic-editing techniques (see CRISPR below) to turn them into cancer cell seek-and-destroy killing machines. Finally, they grow a shitload of them in a lab somewhere and stick them back into the patient. The result: No more cancer. Usually.

It's more complicated than that, and the biggest complication comes from the fact that, in order to remove these T-cells and make new ones that won't be attacked by the old ones, they basically have to wipe out the patients' immune systems. They keep them in isolation for that time, and literally anything that wants to attack them will attack them. They are vulnerable to any opportunistic viruses in their own bodies, plus hospitals aren't exactly winning any awards for being staph infection–free.

But when it works, the results are *amazing*. From start to finish, Emily Whitehead's treatment was about two weeks long. And again, she is not in remission: Her body is permanently ready to destroy any leukemia cells that crop up. She is literally cancer-free.

This means that if you get mono in high school, you will be out for longer *than if you get cancer*. God I love science!

CRISPR

The other day, I was reading *BusinessInsider.com*, a website with only slightly more credibility than *HuffPost* and slightly less than Snapple facts. I found an article from 2012 titled "30 Innovations That Will Change the World." Number one on that list? Edible food packaging that will eliminate plastic waste. Evidently, you'll finish a box of Tic-Tacs, then eat the box. Nowhere on the list? Affordable gene manipulation that will rid the world of all known diseases.

Because it didn't exist then. But it does now, and it is super-amazing and everything they say it is.

Before we go further, let me explain a simple element of gene therapy that most people skip over. Many diseases and disorders are inherited from the parents, and some are mutations that are new to the afflicted generation. But a mutated or faulty genetic sequence doesn't necessarily mean the troublesome genes aren't there; it usually means that, for whatever reason or mistake in the code, the genes don't do their job. This is called gene expression, and it is activated either by routine processes or by reactions to stimuli, like switching from daylight to night, or hormones, or trauma.

Some gene therapy is there to find a different stimulus to get the same result. Not getting the proper chemical changes that help you recuperate during sleep? We'll find something that will stimulate your genes to do the job right.

Other gene therapy actually goes into the genetic code of your DNA and fixes it so that everything is working properly and you even get to pass the corrected code on to your kids. And before CRISPR, this was really just a fantasy.

CRISPR stands for "clustered regularly interspaced short palindromic repeats," which you are immediately allowed to forget because the acronym sounds like a drawer in your refrigerator and is easier to remember. The name describes a kind of molecule mirrored across its length—hence *palindromic*—that exists in single-cell organisms like bacteria that they use in defense from viral infection. (Viruses are basically little DNA-delivery systems to mess you up.)

It is also a process whereby you create customized molecules like this, but they are made in such a way as to be able to very specifically, very accurately edit the genetic code in a DNA base pair. The goal is to fix what's wrong.

My nephew Brady is an infant who suffers from epidermolysis bullosa, or EB, a rare mutation that doesn't allow his skin to produce collagen, the stuff that keeps skin together—and on your body. It's rare and extremely devastating. The skin of these so-called butterfly children blisters and sloughs off with very little provocation, causing scarring that then pulls away more skin as the child grows. At six months, he is covered in scars. Now the skin in his throat does the same thing. His tongue is fused to the floor of his mouth. He is bandaged head to toe and has a feeding tube inserted into his abdomen. But man, that kid is brave. Through the scarring and bandages, you can almost always see a bright little smile.

One of the many problems with a rare disease is there is very little money to cure it. Not enough people have it to be able to raise that money. But the CRISPR technology allows scientists to go after these long-tail disorders, and they are trying to cure it *right now*. And everything they learn about curing this rare disease goes into curing the next one.

So, as far as CRISPR goes, the future can't come soon enough!

AFTERWORD

If you're reading this page, I want to thank you for making it all the way through the book—or at least thanks for skipping to the end in the belief that you were reading a mystery novel, in which case the marketing department of mighty Weldon Owen Publishing thanks you for confusing their awareness campaign for that of a mystery novel and buying the book anyway.

Pretty good-looking book, right?

What I can tell you with absolute certainty at this point is that there are mistakes in this book. And there will be more mistakes as time goes on, due to the nature of the material. It's not that the fundamental physical properties of the universe keep changing; we are laboring under the presupposition that they never change. It is our *perception* of those laws that will continue to change as we come to a deeper understanding of the world around us, the bodies that house us, and the people with whom we connect.

Making simple human connections makes us live longer. That connection can come in many forms, like reading a novel—or even playing a video game. A study a few years back revealed that children with autism showed improvements in socialization after sessions of role-playing video games not only due to the game interactions but also because they had to compete to be good at it, and think like the programmers to advance in the game. But the real stuff is sitting down and having a conversation with someone.

And walking, more than any other form of exercise, is good for brain health. If you read this book closely, you can tell where I had to stop and go take a walk, coming back to the work with a fresh eye and renewed energy. Study after study proves this. Forget your ten-minute strenuous workout or your ultramarathon—those things are great but can cause injury and require focus. Walking is natural. Striding is striving. Get some steps.

There is a reason we call a lot of the facts like those presented in this book "Cocktail-Party Science." It's to show that you can strike up a conversation with anyone, whether it is a new acquaintance or someone who might have gone stale on you, by telling them something you find to be interesting. Immediately, you will know whether they are receptive to the same kind of ideas and if they share your own sense of curiosity. It is my goal, and Kevin's, with this book and the podcast we love to do that you keep your sense of curiosity alive and that you might use something you heard from us that will help you make a human connection.

One last trivia fact for the book—the origin of *trivia*. If you are a *Jeopardy* fan, it can be a big part of your life. But the word itself has come to mean "something unimportant"—or "trivial," if you will. Of course, word origins and meanings can be very different—after all, the latest version of the *Oxford English Dictionary* includes a definition for *literal* that is, essentially, "figurative." But *trivia*, that's another story. It's literally Latin for "three roads," and it comes from the fork at the end of the Appian Way, outside of ancient Rome. The road split into three directions, and there at the crossroads was a kiosk with notices, help-wanted ads, lost-dog bills, political commentary, ads for inns and other services—you name it. It became so crowded that each slip of paper, scratched-on message, or idea did not contribute to a greater understanding, no matter how important it was in itself. Hence *trivia*.

So, long live trivia. Factoids (in the latest, positive sense of the word) that tend to stick to our brains tend to inform us, though sometimes it might take the form of a patchwork quilt, and not a coherent tapestry. These seemingly random ideas can serve as a sort of a crunchy gravel foundation against which we measure and judge new information. And what is old can be new: Growing up, my kids didn't know what the expression "You sound like a broken record" meant, but now that vinyl is back, it makes perfect sense to them.

Thanks for reading. I can't wait to hear what you have to tell me at your next cocktail party.

—*Andy McElfresh*

EDUMACROSSWORD

ACROSS

1 "Night on ___ Mountain"
5 "It's a ___!"
9 Welch role
13 Yemini scratch
18 Pee-pee component
19 Eucalyptus eater
20 13-Across spender
21 "Your wife has ___ angina."
   "I know that, but what's wrong
   with her heart?"
22 Mos Eisley characteristics
25 Hit the imaginary wall?
26 Blue shoe material
27 Certain times
28 "Take your stinking paws off
   me you damned dirty ___!"
29 Way
32 "When the blitzkrieg raged and
   the bodies ___. . ."
34 Rubber tips
38 Cage song, often
39 Lift with tackle
40 A-line wearers
41 Final degree
42 Certain elitist
45 Feather adherer
46 "Our ___"
48 Soprano solo, often
49 Smidge
50 Asked
51 Quint's ship in "Jaws"
52 "Spanish ___"
53 Stopped
55 Article about one who has
   stopped
56 Wool source
58 Tooth issue
59 Lace tip
60 Irish atmosphere
62 Like a wraith
67 Common vow
68 Bygone movie format
70 "Kings of ___"
71 Milk sources
73 God of war
74 States group of South America
77 Washingtons
78 Resistance units
81 Grammys
82 Podcast-cum-so-so-Amazon
   show
83 "Clear the ___!"
84 Leave Out

85 Fall mo.
86 Lunesta, e.g.
90 Manhattan-to-Providence dir.
91 31-Down masterpiece
93 Japanese ingredient
94 Indicate
97 Tore a new one
98 Stonehenge builder, e.g.
99 Pooped
100 Month or term preceder
101 "The end is ___"
102 Gets one's swerve on
103 Dameron portrayer
106 Chisenbop user, literally
112 Seal
113 Like
114 Hose color
115 Life of Riley descriptor
116 Finishes a lawn
117 Give (out)
118 "Incredibles" Edna
119 Tears

DOWN

1 USB, e.g.
2 Lamp type
3 Romanian moolah
4 Plum relative
5 Apres-ski quaff
6 Ecstasy event
7 Actress Larter or Wong
8 Bub
9 Took in
10 Kayak cousin
11 "___ of the father"
12 Put before a jury
13 Infamous L.A. police division
14 Least friendly
15 KFC parent company
16 Etched
17 Screen type
19 Tollings
23 Certain glow
24 Car agreement
28 Ark destination
29 Fruit with a big, flat seed
30 Fragrant resin
31 91-Across auteur
32 Hagia ___
33 Mineral springs rock
34 Gives off
35 Radical opponent
36 Prepped

37 Quick way to end an aria
39 The "B" in FBI
40 Comics newspaperbird
43 Soapstones
44 Rock-sitting chanteuse
47 Hoedown participant
50 Fen
52 Enthusiast
53 Not just think about
54 Comparison word
57 "Yellow matter custard,"
   per a Beatles song
58 One way to spend a semester
59 It can come in a yard
60 Trifecta's place
61 Korn cover "___ My Eye"
63 Censor's tool
64 Watch, in a way
65 Betsy Wetsy maker
66 Spots
69 Kind of curve
72 Pooh friend
74 Joined forces
75 Very much
76 Urn bottom
77 Invading weed in
   North America
79 Prepare a meat pie
80 Peel's avenging partner
83 White tomb site
86 Routines
87 Hopeful Lennon song
88 "A ___ at the Opera"
89 Bean, e.g.
92 Husk-wrapped meal
95 Louis-Dreyfus vehicle
96 Like the DMV, vis-a-vis
   licenses
98 Scottish investment firm
99 Altar-nate ceremony?
101 Its headquarters are in
   Beaverton, OR
102 Gulf War missile
103 Whack
104 Presidential hashtag
105 Leo mo.
106 Hoover, e.g.
107 Cash spot
108 Vietnamese neighbor
109 ___ chi
110 With it, you might see
   dead people
111 "___ ipsa loquitor"

ABOUT THE AUTHORS

Andy McElfresh is an entertainment producer and the co-host of the *Edumacation* podcast. He was formerly a writer and segment director for *The Tonight Show with Jay Leno*, wrote the screenplay for *White Chicks*, created Nickelodeon's *Rocket Power*, wrote for MTV, Comedy Central, and the *Keenan Ivory Wayans Show*, and was the first employee at the Zagat Survey. He lives in Los Angeles.

Kevin Smith is a *New York Times* bestselling author and the co-host of the *Edumacation* podcast. He wrote and directed *Clerks* and *Chasing Amy*, directed episodes of *The Flash*, *Supergirl*, and *The Goldbergs*, and produced the Oscar-winning film *Good Will Hunting*. He lives in Los Angeles.

Kelsey Dake is an award-winning illustrator whose work has appeared in *The Atlantic*, *New York Times*, *GQ*, and *Wired*, and was named an ADC Young Gun and one of *Print* magazine's Top 20 Under 30. She lives in Phoenix.

EDUMAANSWERS

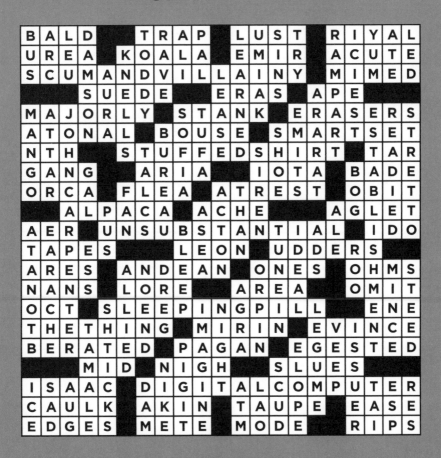

BALD · TRAP · LUST · RIYAL
UREA · KOALA · EMIR · ACUTE
SCUMANDVILLAINY · MIMED
· SUEDE · ERAS · APE
MAJORLY · STANK · ERASERS
ATONAL · BOUSE · SMARTSET
NTH · STUFFEDSHIRT · TAR
GANG · ARIA · IOTA · BADE
ORCA · FLEA · ATREST · OBIT
· ALPACA · ACHE · AGLET
AER · UNSUBSTANTIAL · IDO
TAPES · LEON · UDDERS
ARES · ANDEAN · ONES · OHMS
NANS · LORE · AREA · OMIT
OCT · SLEEPINGPILL · ENE
THETHING · MIRIN · EVINCE
BERATED · PAGAN · EGESTED
· MID · NIGH · SLUES
ISAAC · DIGITALCOMPUTER
CAULK · AKIN · TAUPE · EASE
EDGES · METE · MODE · RIPS

President & Publisher Roger Shaw

SVP, Sales & Marketing Amy Kaneko

Associate Publisher Mariah Bear

Acquisitions Editor Kevin Toyama

Creative Director Kelly Booth

Art Director Allister Fein

Production Designer Howie Severson

Production Director Michelle Duggan

Production Manager Sam Bissell

Imaging Manager Don Hill

Weldon Owen would like to thank Mark Nichol for his copyediting expertise, Carrie Shepherd for her careful proofreading skills, and Scott Thorpe and Shannon Mitchner for their invaluable teamwork.

ISBN: 978-1-68188-301-4

Printed in China

Cover Photo by Allan Amato
Designed by Laura Bagnato for MacFadden & Thorpe

10 9 8 7 6 5 4 3 2 1
2018 2019 2020 2021

Weldon Owen
1045 Sansome Street, Suite 100
San Francisco, CA 94111
www.weldonowen.com

Weldon Owen is a division of Bonnier Publishing USA (www.bonnierpublishingusa.com)